Debates in Design and Technology Education

Design and technology has long held a controversial place on the school curriculum, with some arguing that it shouldn't be there at all. This book presents and questions considered arguments and judgements, and explores the major issues that all design and technology teachers encounter in their daily professional lives. In exploring some of the key debates, it encourages critical reflection and aims to stimulate both novice and experienced teachers to think more deeply about their practice, and link research and evidence to what they have observed in schools.

Written by expert design and technology education professionals, chapters tackle established and contemporary issues, enabling you to reach informed judgements and argue your point of view with deeper theoretical knowledge and understanding. Debates covered include:

- What is the purpose of design and technology?
- Is it a vocational or academic subject?
- What is the place of design and technology within the STEM agenda?
- What knowledge and skills do teachers really need?
- What does the design and technology gender divide mean for schools and pupils?
- Is it a 'creative' subject?
- What is the future for design and technology?

With its combination of expert opinion and fresh insight, *Debates in Design and Technology Education* is the ideal companion for any student or practising teacher engaged in initial training, continuing professional development or Masters level study.

Gwyneth Owen-Jackson is Senior Lecturer at the Open University, UK.

Debates in Subject Teaching Series
Series edited by: Susan Capel, Jon Davison, James Arthur, John Moss

The **Debates in Subject Teaching Series** is a sequel to the popular **Issues in Subject Teaching Series**, originally published by Routledge between 1999 and 2003. Each title presents high-quality material, specially commissioned to stimulate teachers engaged in initial training, continuing professional development and Masters'-level study to think more deeply about their practice and link research and evidence to what they have observed in schools. By providing up-to-date, comprehensive coverage the titles in the **Debates in Subject Teaching Series** support teachers in reaching their own informed judgements, enabling them to discuss and argue their point of view with deeper theoretical knowledge and understanding.

Titles in the series:

Debates in History Teaching
Edited by Ian Davies

Debates in English Teaching
Edited by Jon Davison, Caroline Daly and John Moss

Debates in Religious Education
Edited by Philip Barnes

Debates in Citizenship Education
Edited by James Arthur and Hilary Cremin

Debates in Art and Design Education
Edited by Lesley Burgess and Nicholas Addison

Debates in Music Teaching
Edited by Chris Philpott and Gary Spruce

Debates in Physical Education
Edited by Susan Capel and Margaret Whitehead

Debates in Geography Education
Edited by David Lambert and Mark Jones

Debates in Design and Technology Education
Edited by Gwyneth Owen-Jackson

Debates in Design and Technology Education

Edited by
Gwyneth Owen-Jackson

Routledge
Taylor & Francis Group

LONDON AND NEW YORK

First published 2013
by Routledge
2 Park Square, Milton Park, Abingdon, Oxon OX14 4RN

Simultaneously published in the USA and Canada
by Routledge
711 Third Avenue, New York, NY 10017

Routledge is an imprint of the Taylor & Francis Group, an informa business

© 2013 Gwyneth Owen-Jackson

The right of the editor to be identified as the author of the editorial material, and of the authors for their individual chapters, has been asserted in accordance with sections 77 and 78 of the Copyright, Designs and Patents Act 1988.

All rights reserved. No part of this book may be reprinted or reproduced or utilised in any form or by any electronic, mechanical, or other means, now known or hereafter invented, including photocopying and recording, or in any information storage or retrieval system, without permission in writing from the publishers.

Trademark notice: Product or corporate names may be trademarks or registered trademarks, and are used only for identification and explanation without intent to infringe.

British Library Cataloguing in Publication Data
A catalogue record for this book is available from the British Library

Library of Congress Cataloging in Publication Data
Debates in design and technology education / edited by Gwyneth Owen-Jackson.
pages cm
Includes bibliographical references and index.
ISBN 978-0-415-68904-5 -- ISBN 978-0-415-68905-2 --
ISBN 978-0-203-51949-3 1. Technology--Study and teaching.
2. Design--Study and teaching. I. Owen-Jackson, Gwyneth, 1956-
T65.D36 2013
607.1--dc23
2012043949

ISBN: 978-0-415-68904-5 (hbk)
ISBN: 978-0-415-68905-2 (pbk)
ISBN: 978-0-203-51949-3 (ebk)

Typeset in Galliard
by Saxon Graphics Ltd, Derby

Printed and bound in Great Britain by MPG Printgroup

Contents

List of figures and tables	vii
List of contributors	ix
Acknowledgements	xiii
List of abbreviations	xv
Introduction to the series	xvii

Introduction   1

PART I
Setting the scene   5

1. Government policies and design and technology education   7
   DANIEL WAKEFIELD AND GWYNETH OWEN-JACKSON

2. Developments in teaching design and technology   21
   DANIEL WAKEFIELD

3. International perspectives on technology education   31
   FRANK BANKS AND P. JOHN WILLIAMS

PART II
Debates about design and technology   49

4. Why is transition from primary to secondary school so difficult?   51
   CATHY GROWNEY

5. Is design and technology about making or knowing?   64
   MIKE MARTIN AND GWYNETH OWEN-JACKSON

6 Design and technology education: vocational or academic?
   A case of yin and yang   74
   GARY O'SULLIVAN

7 What makes a good technology teacher?   86
   NIGEL ZANKER AND GWYNETH OWEN-JACKSON

PART III
Debates within design and technology   99

8 Does food fit in design and technology?   101
   SUZANNE LAWSON

9 Textiles: design and technology or art?   115
   CHRIS HUGHES AND DAVID WOOFF

10 Using technology in design and technology   125
   ALISON HARDY AND SARAH DAVIES

11 STEM: opportunity, opposition or just good design
   and technology?   139
   TONY COWELL

PART IV
Debates about teaching design and technology   151

12 The (continuing) gender debate   153
   DAWNE BELL, CHRIS HUGHES AND GWYNETH OWEN-JACKSON

13 Creativity for a new generation   166
   DAVID SPENDLOVE AND ALASTAIR WELLS

14 Assessment questions   180
   DAVID WOOFF, DAWNE BELL AND GWYNETH OWEN-JACKSON

Endpiece   193
Index   197

List of figures and tables

Figures

7.1	A visual tool for describing teachers' professional knowledge	88
7.2	Knowledge for Teaching (adapted from TDA 2007)	88
10.1	Model of a teacher's personal philosophy of education	134

Tables

1.1	Developments in politics and curriculum since 1976	9
1.2	D&T programmes of study 1990	13
3.1	Technology topics in Japan	41
4.1	Year 6 and Year 7 differences	57
5.1	Skills and knowledge in the teaching of design and technology	66
8.1	Home economics examination entries classified by gender	104
8.2	Food technology examination entries classified by gender	105
10.1	Scope of technology in D&T considered in this chapter	126
10.2	Stages of teaching and learning with technology in D&T	132
10.3	Cost of ICT equipment	135

List of contributors

Frank Banks is Professor of Teacher Education and Director for International Development in Teacher Education at the Open University. An ex-schoolteacher, with much experience in teacher education, he is responsible for a number of teacher professional development projects in India, Bangladesh and sub-Saharan Africa.

Dawne Bell has over 15 years' teaching experience, having moved into education from industry, and at Edge Hill has responsibility for work across a range of ITT programmes. Specific to D&T, Dawne's current role is course leader for the D&T Flexible PGCE, but she holds overall responsibility for the co-ordination and development of Art, Design and Technology across the Faculty of Education. She is also a SOLSTICE Learning and Teaching Fellow.

Tony Cowell qualified to teach in 1990, having originally been an engineer, and taught D&T in South Yorkshire schools for 18 years. Tony was appointed Senior Lecturer at Sheffield Hallam University Centre for Design and Technology Education in 2008 and received a SHU inspirational teaching award in November 2011.

Sarah Davies is the programme leader for post-graduate Secondary Education Initial Teacher Education (ITE) courses at Nottingham Trent University. An experienced advanced skills teacher, Sarah joined NTU in 2007 after 10 years working in secondary education within the East Midlands. Her research centres around teacher professional development and the use of digital technology within the classroom.

Cathy Growney is the 'Technology Outreach Teacher' at a secondary school in West Berkshire with responsibility for D&T partnerships with 10 primary schools. She has taught on ITE courses for primary and secondary teachers. In her current work, Cathy collaborates with primary and secondary teachers and teaches D&T to pupils aged 4–18. Her D&T research includes primary technology education in England; primary–secondary transition; global dimensions in D&T; and comparative research in D&T.

Alison Hardy is the programme leader for the undergraduate ITE Design and Technology course at Nottingham Trent University. An experienced teacher and head of department, Alison has also worked in the further education sector managing school/college partnerships. Her research includes the value of D&T, technology in D&T and ITE.

Chris Hughes was formally Senior Lecturer in D&T at Edge Hill University with responsibility for the undergraduate course in D&T with QTS. He has worked at all levels of education including secondary schools, a sixth form college and universities. Chris has wide experience in textiles, apparel applications and general textile manufacturing. His research interests include student teachers' perceptions of their practice, D&T pedagogy and the role of D&T in the wider cultural context.

Suzanne Lawson is Senior Lecturer in Design and Technology at Birmingham City University and route leader for the PGCE in Design and Technology. Research interests include food education, children's perceptions of food technology, and CPD courses for textiles teachers.

Mike Martin has been involved in design and technology for 20 years, teaching, providing professional development, lecturing and external examining across eight ITE institutions. He is known for his work on values, and currently subject knowledge, presenting regularly at national and international conferences.

Gary O'Sullivan is a Senior Lecturer in Technology Education at Massey University in New Zealand, where he is responsible for primary and secondary teacher education programmes. He has researched, published and presented both nationally and internationally for the technology community.

Gwyneth Owen-Jackson has been involved with D&T education for over 20 years, as a secondary teacher, with government organisations in an advisory capacity, with the D&T Association, and as a teacher educator. She has edited several books used in teacher education programmes. Research interests include the professional identity of D&T (student) teachers and primary education in Bosnia and Herzegovina.

David Spendlove is the Director of Secondary Teacher Training at the University of Manchester. He is co-editor of *Design and Technology Education: An International Journal* and has written extensively about creativity, gender, emotion and assessment.

Daniel Wakefield has been teaching for 10 years and is currently Teacher in Charge of D&T at Birkenhead High School Academy girls school (3–19). He is particularly interested in the development of the subject, classroom practice and leadership of D&T within schools.

Alastair Wells is Director of Unlimited Paenga Tawhiti, Christchurch, New Zealand. He has worked in education at all levels and research interests include: design, creativity, cognition, innovative pedagogies, technology, educational psychology and educational environments.

P. John Williams is an Associate Professor and Director of the Centre for Science and Technology Education Research at the University of Waikato in New Zealand, where he teaches and supervises research students in technology education. Apart from New Zealand, he has worked and studied in a number of African and Indian Ocean countries and in China, Australia and the United States.

David Wooff spent over 10 years teaching in secondary education, before moving into a career in ITT. Initially joining Edge Hill University as the undergraduate course leader for full-time and part-time D&T ITE courses, he has recently moved to a managerial role within the Faculty of Education. David is now Head of ITT Partnership at Edge Hill University.

Nigel Zanker is PGCE/MSc Programme Director for the Design School at Loughborough University. His experience includes Ofsted Inspection and Electronics in School Strategy training. His principal areas of research are pedagogic subject knowledge development in design and technology through action research. He is widely published in design education.

Acknowledgements

In these times where everyone is working harder and longer than ever before, and where the future of design and technology and teacher education is uncertain, thank you to all the authors for the time that they have taken to make their contributions to this book. Thank you, too, to the schools, teachers and students who have allowed their work to be discussed.

Acknowledgement is given to Frank Banks for his permission to use Figure 7.1 and to the Design and Technology Association for permission to use the tables in Chapter 8.

Thank you to my colleagues at the Open University who have given me the time and space to edit this book, especially given all the demands being made on their time. Thank you to my editors, who have shown patience and given me much support.

Thank you to Jennifer for her patience, understanding and practical support throughout the production of this book.

Chapter 4

My thanks to all those colleagues and pupils who have contributed their thoughts and evidence to this chapter, especially to Lynette Bodsworth, Liz James and Barbara Lowe.

List of abbreviations

CDT	craft, design and technology
D&T	design and technology
DCSF	Department for Children, Schools and Families
DES	Department of Education and Science
DfE	Department for Education
DfES	Department for Education and Science
ICT	information and communication technology
NC	National Curriculum
NCC	National Curriculum Council
QCA	Qualifications and Curriculum Authority
STEM	science, technology, engineering and mathematics
TA	Teaching Agency
TDA	Training and Development Agency

Introduction to the series

This book, *Debates in Design and Technology Education*, is one of a series of books entitled *Debates in Subject Teaching*. The series has been designed to engage with a wide range of debates related to subject teaching. Unquestionably, debates vary among the subjects, but may include, for example, issues that:

- impact on initial teacher education in the subject
- are addressed in the classroom through the teaching of the subject
- are related to the content of the subject and its definition
- are related to subject pedagogy
- are connected with the relationship between the subject and broader educational aims and objectives in society, and the philosophy and sociology of education
- are related to the development of the subject and its future in the twenty-first century.

Consequently, each book presents key debates that subject teachers should understand, reflect on and engage in as part of their professional development. Chapters have been designed to highlight major questions and to consider the evidence from research and practice in order to find possible answers. Some subject books or chapters offer at least one solution or a view of the ways forward, whereas others provide alternative views and leave readers to identify their own solution or view of the ways forward. The editors expect readers will want to pursue the issues raised and so chapters include questions for further debate and suggestions for further reading. Debates covered in the series will provide the basis for discussion in university subject seminars or as topics for assignments or classroom research. The books have been written for all those with a professional interest in their subject and, in particular: student teachers learning to teach the subject in secondary or primary school; newly qualified teachers; teachers undertaking study at Masters level; teachers with a subject co-ordination or leadership role and those preparing for such responsibility; as well as mentors, university tutors, CPD organisers and advisers of the aforementioned groups.

Books in the series have a cross-phase dimension, because the editors believe that it is important for teachers in the primary, secondary and post-16 phases to look at subject teaching holistically, particularly in order to provide for continuity and progression, but also to increase their understanding of how children and young people learn. The balance of chapters that have a cross-phase relevance varies according to the issues relevant to different subjects. However, no matter where the emphasis is, the authors have drawn out the relevance of their topic to the whole of each book's intended audience.

Because of the range of the series, both in terms of the issues covered and its cross-phase concern, each book is an edited collection. Editors have commissioned new writing from experts on particular issues, who, collectively, represent many different perspectives on subject teaching. Readers should not expect a book in this series to cover the entire range of debates relevant to the subject or to offer a completely unified view of subject teaching; neither should they expect that every debate will be dealt with discretely or that all aspects of a debate will be covered. Part of what each book in this series offers to readers is the opportunity to explore the interrelationships between positions in debates and, indeed, among the debates themselves, by identifying the overlapping concerns and competing arguments that are woven through the text.

The editors are aware that many initiatives in subject teaching continue to originate from the centre and that teachers have decreasing control of subject content, pedagogy and assessment strategies. The editors strongly believe that for teaching to remain properly a vocation and a profession, teachers must be invited to be part of a creative and critical dialogue about subject teaching, and should be encouraged to reflect, criticise, problem solve and innovate. This series is intended to provide teachers with a stimulus for democratic involvement in the development of the discourse of subject teaching.

Susan Capel, Jon Davison, James Arthur and John Moss
December 2010

Introduction

Design and technology is a relative newcomer to the school curriculum compared to other subjects, Williams (1961: 130) notes that 'the quadrivium of music, arithmetic, geometry, and astronomy goes back to at least the fifth century'. Design and technology, in contrast, has been on the curriculum in the UK for just over 20 years and in some countries for less.

This short history is an influential factor in many of the debates aired in this book. Unlike many of its curriculum contemporaries, D&T does not have a long tradition on which to draw for its identity and its pedagogy. That is not to say that it has no history. As several authors note here (Wakefield and Owen-Jackson, O'Sullivan) there are antecedents to D&T; in the mid-sixteenth century schools were established to provide apprenticeships in trades and crafts and in the early nineteenth century girls were taught spinning, sewing and how to cook (Gillard 2011). Design and technology, however, is not the same as these. Yet, as some of the discussions show, it is easier to say what D&T is *not* than what it *is*.

The first four chapters in this book, in Part 1, 'Setting the scene', draw heavily on this historical background. In Chapter 1, Daniel Wakefield and Gwyneth Owen-Jackson provide the history of D&T in the UK curriculum, as this is where the subject, as we currently know it, began. Although some aspects of D&T, mainly the practical elements of 'making', had been on the curriculum for some time and craft, design and technology (CDT) preceded it and contributed much to the development of D&T, it was the UK National Curriculum of 1988 that conceived and introduced the subject as we now know it. Chapter 1 charts the development of D&T from its inception through to the present day, highlighting the source of some of the current and ongoing discussions. The chapter notes that the precedents of the subject and contemporary social influences on education meant that boys and girls experienced different subjects, a division that still exists in D&T today and which is discussed in detail by Dawne Bell, Chris Hughes and Gwyneth Owen-Jackson in Chapter 12. It was political influences, however, that led to the introduction of design and technology to the curriculum for all pupils, and political influences that led to the lack of clarity surrounding the nature and purpose of this new subject, a debate explored further by Mike Martin in Chapter 5 and Gary O'Sullivan in Chapter 6.

The importance of the discussion of the conception, introduction and implementation of D&T is demonstrated by the fact that many of the chapters in this book refer to the history and development of the subject in order to locate the contemporary discussion; it is important to understand how we got to where we are. Wakefield continues this historical look at the subject in Chapter 2, where he considers how the developments in D&T discussed in Chapter 1 have influenced its teaching. Drawing on a sample of projects and initiatives to illustrate, the chapter discusses how teachers' practice has changed and continues to change.

The introduction of D&T in the UK was influential in many other countries. They all had their own traditions on which to draw, the Sloyd tradition of Sweden and the indigenous technologies of African countries, but many of these have developed in surprisingly similar ways. In Chapter 3, Frank Banks and P. John Williams look at the curricula of D&T, or technology, in a number of countries. This highlights both similarities and differences and contributes to the debate about the nature and purpose of D&T education.

Part 2, 'Debates about design and technology', discusses some of the broader issues within the subject. Chapter 4 examines pupils' experience of D&T as they move from primary to secondary school. For many pupils, this is a time of both excitement and anxiety, but in D&T there seems to be disruption in their learning, Cathy Growney considers why this is and what might be done to alleviate the difficulties. Chapters 5 and 6 each focuses in more detail on the content of D&T. In 2001, writing about D&T on the National Curriculum, Moon noted that 'Early debates centred on the balance between the theoretical and practical aspects of the subject' (Moon 2001: 59). In Chapter 5, Mike Martin and Gwyneth Owen-Jackson return to this long-standing debate on whether the focus of D&T should be on developing pupils' practical skills or their knowledge. Another debate of long standing is whether D&T is a vocational subject or one that contributes to pupils' general education and Gary O'Sullivan addresses this in Chapter 6. The content and purpose of D&T on the school curriculum, discussed in these chapters, informs what teachers need to know and do, which is the focus of Chapter 7, when Nigel Zanker and Gwyneth Owen-Jackson discuss 'What makes a good technology teacher?'.

There are also ongoing discussions within the D&T community about various aspects of the subject itself and these are approached in Part 3, 'Debates within design and technology'. The debates here are about whether or not food fits within D&T, discussed by Suzanne Lawson in Chapter 8, whether textiles belongs in D&T or art and design, asked by Chris Hughes and David Wooff in Chapter 9. The same considerations are not given to resistant materials or electronics as these are understood by many to constitute design and technology, and their place is not contested. This could, of course, form its own debate – the contribution of resistant materials and electronics to pupils' learning in D&T – but lack of space prevented its inclusion. However, D&T is also often misunderstood as being solely about 'technology', by which is often meant computer technology. In Chapter 10, Alison Hardy and Sarah Davies consider what use do we, and should we, make of computer technology in D&T. Finally in this part, in Chapter 11, Tony Cowell addresses the question of the position of D&T within the science, technology, engineering and mathematics (STEM) agenda. STEM is an issue in many countries, driven by concerns about future technological and economic developments. In these broader discussions there has been much emphasis on science and mathematics but little attention, until recently, paid to (design and) technology and engineering. Cowell shows the importance of making sure that T and E are discussed equally with S and M.

Part 4, 'Debates about teaching design and technology', returns to broader discussions within the subject. Dawne Bell, Chris Hughes and Gwyneth Owen-Jackson look at the continuing gender divide in D&T (Chapter 12), asking why it still exists and what might be done to address it. In Chapter 13, David Spendlove and Alistair Wells present an interesting discussion about the place of creativity within D&T, highlighting its apparent lack, suggesting new ways in which it might be considered and stimulating further debate. Chapter 14, by David Wooff, Dawne Bell and Gwyneth Owen-Jackson, brings together much of what has gone before by debating some of the issues around assessment in D&T. Effective assessment requires that what is to be assessed is known and understood and these authors ask if that is the case in D&T. They also ask if the assessment methods we use are as valid and reliable as we think they are. These are important questions and in order to be able to answer them we need to have a clear understanding of the subject and how pupils learn it – the concern of most of this book.

The Endpiece summarises the discussions in the book and tries to look forward to what future debates and discussions will ensue.

It has to be noted that the debates and discussions presented here represent only a selection of those that could have been included. It is acknowledged that some have been missed, for example around the place of 'design' in D&T; is design a general skill or is it different in the different areas of D&T? The place of values in D&T is not fully explored, neither are the issues around the teaching about sustainability within D&T. It is not because these debates are not important, but constraints on space meant that choices had to be made. It is hoped that debate on these other important issues will continue alongside the debates presented here.

Debate is healthy, it indicates an interest and a passion in the subject and can contribute to moving the subject forward. In contrast, it can also lead to fragmentation and polarisation. Both of these have been evident in D&T over the last 25 years. This book aims to present some of the main issues in the subject, many of which have resonance in countries across the world, in the hope that students and teachers of the subject will continue to show interest and passion, continue to debate and continue to develop the subject.

References

Gillard, D. (2011) *Education in England: a brief history,* www.educationengland.org.uk/history (accessed 8 March 2012).

Moon, B. (2001) *A guide to the National Curriculum*, 4th edn, Oxford: Oxford University Press.

Williams, R. (1961) *The long revolution*, London: Chatto & Windus.

Part I

Setting the scene

Chapter 1

Government policies and design and technology education

Daniel Wakefield and Gwyneth Owen-Jackson

Introduction

Educational historians will point out many significant developments, policies and legislation that have shaped the English education system. However, in recent history the 1988 Educational Reform Act, and specifically the introduction of the National Curriculum for England and Wales, had a major impact on what and how pupils are taught. The 1988 Act introduced Design & Technology (D&T) into the National Curriculum and this chapter considers the development of the subject in the UK since then. It considers the influences of the Department for Education (DfE), the Design and Technology Working Group, the National Curriculum Council and teachers in shaping the curriculum and influencing the perception of D&T, and the impact that this has had.

In order to explore the extent to which the National Curriculum has influenced the subject as it is today we need to consider:

- where it all began
- the introduction of the National Curriculum
- National Curriculum developments.

Where it all began

A review of the history of (design and) technology in the UK reveals that controversy and division has been present throughout. Early education in Britain, as in many other countries, was provided by the church and was an academic preparation for the upper classes. Trade and craft skills were learnt through apprenticeship and it was not until the mid-sixteenth century that these became a formal part of the education system (Gillard 2011). The Industrial Revolution in Britain, which took place between 1750–1850, led to massive social and political change, which included the expansion of basic education to those who had previously not had access. During the Industrial Revolution major design, industrial and manufacturing advancements made Britain a leading industrial nation but, by the end of the nineteenth century, this was no longer the case and Britain lagged behind many of its competitors. This could be attributed partly to the lack of attention to technical education for its young people as, during the same period, countries across Europe had been investing both in industry and education.

Although there is evidence of technological activities in school before 1902, that is the date at which 'manual work' was introduced to English state schools and many of our issues with design and technology education may be traced back to this era. Woodwork was a mandatory subject for boys while 'housewifery' was the option for girls (Mulberg 1992). The development of manual work on the curriculum in England, and to the same extent the American elementary system, had a utilitarian approach 'as a pre-vocational training for the rapidly growing ranks of manual and domestic workers' (Eggleston 1976: 5).

In the early twentieth century, manufacturing, industry and domestic life were very different from those of today. Highly skilled craftsmen and well-trained, resourceful housewives were the ideals of the day. Industry and the factory workshop were predominantly the domain of the male, so exposing boys to industrial practices and techniques at school allowed them to develop skills and attitudes for the workplace. Little emphasis was placed on the challenge of new design:

> The things boys made solved no problems. There was no challenge to think either why or how. Shape, size, construction, materials, tools, processes were all taken for granted, and as a result much of the educational value of the work was lost. There was little connection between the crafts and the technical drawing intended to be the language of expression.
>
> (Kingsland 1969: 11)

The Spens Report of 1938 introduced the idea of 'technical' schools alongside grammar schools and this was enacted in the 1944 Education Act, which introduced into Britain the 'tripartite' system of education. This provided grammar schools for academic pupils, technical schools to teach mechanical and engineering skills and secondary modern schools for the academically less able. For some reason the technical schools, which would have been the ideal place for technology education, failed to properly establish themselves and a great opportunity was missed to make a technical education a viable alternative for pupils.

Over the course of the twentieth century, through various acts of parliament, the government became increasingly involved in educational provision, leading to the 1988 Education Act, which introduced the first National Curriculum in the UK.

Introduction of the National Curriculum

Educational developments do not occur in a vacuum, they are part of the political and social context of the time. Table 1.1 provides an outline of some of the key

Table 1.1 Developments in politics and curriculum since 1976

Year	Event
1976	James Callaghan (Prime Minister) speech at Ruskin College introduced the 'Great Debate' about the nature and purpose of the school curriculum and the need for a basic curriculum followed by all
1987	Foundation for the National Curriculum outlined by the Conservative government
1988	Design and Technology Working Group Interim Report published
1989	Final report by working group presented to DES
1990	First technology curriculum implemented (note change of name)
1991	Guidance and support material published to support teachers
1993	Full National Curriculum review for all subjects (lead by Sir Ron Dearing) and redeployment of subject working groups
1994	Statutory Technology at Key Stage 4 ended NC Review (the Dearing Report) published Design and technology identified as a subject in its own right
1995	New orders implemented at Key Stages 1, 2 and 3
1996	New orders implemented at Key Stage 4, starting with Year 10 (first year of upper secondary school)
1997	New orders implemented at Key Stage 4, Year 11 Labour Party victory at general election (May) Excellence in Schools white paper published
1998	National Literacy Strategy introduced (in primary schools, intended to raise standards)
1999	Full National Curriculum review in England All subject descriptions published for the first time National Numeracy Strategy introduced
2000	New National Curriculum launched Launch of specials schools Learning Skills Act allowed city technology colleges to become city academies
2004	Five-year Strategy for Children and Learners allowing all schools to become specialists and expanding academy schools 14–19 Curriculum and Qualifications Reform report published suggesting changes to traditional qualifications
2005	Labour wins third term in office
2006	Cambridge primary review launched
2007	QCA consultation over new curriculum for Key Stages 3 and 4
2008	Diploma qualification launched (in particular engineering, manufacturing and construction)
2009	Cambridge primary review and the Rose Report (primary) published
2010	New government elected, coalition between Conservative Party and the Liberal Democrats
2011	Call for evidence towards National Curriculum review (first teaching 2014) Introduction of the English baccalaureate (Ebacc)
2012	Design and technology on the National Curriculum in England to be reviewed

events that have shaped today's curriculum. In order to analyse the changes in government thinking and legislation regarding D&T, it is important to appreciate some of the wider curriculum developments that have taken place and the context in which these occurred.

In 1976 the then Prime Minister, James Callaghan, made a speech at Ruskin College in which he said, among other things, that consideration should be given to a basic curriculum to be agreed for all pupils. This led to a series of discussions and documents and in 1987 the Secretary of State for Education, Kenneth Baker, launched foundations for a National Curriculum as the keystone of the Education Reform Act 1988. The National Curriculum meant that, for the first time in England and Wales, all pupils would be taught (broadly) the same skills and knowledge in 11 subject areas and would be assessed by the same criteria.

The Department of Education and Science (DES) commissioned subject-specific working groups to produce a series of recommendations for subject content, programmes of study, and assessment criteria (attainment targets) (Penfold 1988; Norman, 1990).

The Design and Technology Working Group

The D&T Working Group, led by Lady Parkes, included those with backgrounds in CDT, home economics, vocational studies, science, economics, business studies and information technology (Harris and Wilson 2003). Terms of reference were issued to the Group by the DES, which outlined their duties, principally to develop clear objectives, curriculum content and attainment targets.

The terms of reference make interesting reading; point 6 under the heading of 'approach' identified a number of areas for the Working Group to focus on. They were encouraged to think of the subject as an opportunity for pupils to design and make useful objects. A list of suggested materials was identified and the need to develop pupils' experiences of craft and the world of work were also suggested. The terms highlighted the implicit link between design and technology and information technology but also links with mathematics, science and several other subjects. However, there was a suggestion that technology did not need to become a discrete subject of its own and that many of its skills and knowledge bases could be identified, even located, within other areas of the curriculum. This brings into question immediately what would be the position of technology on the curriculum.

Initially, the task of the Working Group was to develop the curriculum for pupils aged 11–16, with the responsibility for the primary technology curriculum resting with the Science Working Group (McCormick 2002). This was an early signal to science as a senior partner to technology (Penfold 1988). Science was to be a core subject within the curriculum and initial discussions had suggested aspects of technology education could be taught within the science curriculum. However, following an interim report from the Science Working Group, the responsibility for the technology at primary level was transferred to the Design and Technology Working Group. Again it is worth questioning the position of design and technology in the National Curriculum; was it to be the supportive senior party for the rapidly growing information technology department or the junior partner to science?

The Working Group produced a set of proposals that included a rationale of why the subject should be taught. This suggested that design and technology should allow pupils to meet the needs of the twenty-first century and engage in design, investigation and appraisal activities to acquire knowledge. Practical experiences were identified as the main process by which to develop understanding, but there was little emphasis on developing products. The Working Group also proposed five attainment targets:

AT1   Explore and investigate contexts for design and technology capabilities
AT2   Formulate proposals and choose a design for development
AT3   Develop the design and plan for the making of an artefact or system
AT4   Make artefacts and systems
AT5   Appraise the processes, outcomes and effects of design and technological activities.

Questions can be asked about how the initial DES terms of reference were influential in shaping the design and technology proposals. Although a structure was needed for the Working Group, the terms of reference suggested material areas and approaches that were traditional and heavily influenced by CDT. By the same token, there was a drive towards it becoming a more open-ended subject, a subject with roots in many others and its own unique features.

Already, it is possible to see the potential for a clash of identity; is D&T to remain as a discrete subject, will it become a 'science and technology' curriculum or will the traditional making skills continue, with some designing added for modernity? Does this make the position of technology on the curriculum any clearer?

National Curriculum Council

The National Curriculum Council (NCC) was established by the Education Reform Act 1988 to review the work of the subject working groups and produce final consultation documents. The working groups had included experts and practitioners from within the subject but the NCC was less independent from government and, on face value, had little connection with the subject areas (McCormick 2002).

Following a short period of consultation, the technology report was published, and met with mixed reactions. Much of the 'new' subject content was already present in CDT and home economics, but it also drew on elements from art and business studies (Norman 1990). One aspect that caused some consternation in the teaching profession was the apparent loss of CDT and home economics from the school curriculum. Other subjects had seemingly survived but there was a perception that 'technology' was a new subject rather than a development from existing ones. Although this may seem to be a matter of semantics for some, for others it was a major change for the subject.

One noticeable difference between the proposals from the Working Group and the final report was the reduction in the number of attainment targets, from five to four:

AT1   Identify needs and opportunities
AT2   Generate a design proposal
AT3   Planning and marketing
AT4   Appraising.

It was argued that, in reducing the attainment targets, more emphasis had been placed on 'wealth creation and enterprise' (McCormick 2002: 38).

Technology was seen as a new, interdisciplinary subject, a fusion of CDT and home economics with close links to science and mathematics (Eggleston 1996) which led to initial concerns over the practical element of the subjects. Nearly 10 years earlier (1981) the DES had written that CDT and home economics 'make a particular contribution to the acquisition of physical and practical skills which are essential complement of the pupil's intellectual and personal development' (DES 1981: 7).

So, from the outset there were concerns within the technology community over the appropriateness of 'technology' as a subject, the proposed content and its place on the curriculum.

With the implementation of the National Curriculum in 1990, schools in England and Wales became the first in the word to offer technology as a compulsory subject from the ages of 5 to 16 (Kimbell and Perry 2001). However, the implementation and sustainability of the subject was no easy feat.

Technology was described as a subject that would incorporate and expand on work currently being undertaken in art, business studies, CDT, home economics and information technology. It would require a co-ordinated approach among those departments and teachers to help pupils develop their understanding of the 'significance of technology to the economy and life' (NCC 1991: 1). It had also been recognised that in technology 'the capability to investigate, design, make and appraise' was as important as the 'acquisition of knowledge' (DES/WO 1989: 1). The first NC curriculum for technology was, however, packed with information and long lists of content for designing and making. Table 1.2 shows that, although at each stage of schooling the main aspects to be taught remained the same, the volume of content (shown by the number of statements under each heading) varied. The sheer volume of information and content, and the need to work with teachers from other departments, meant it was not easy for teachers to understand or implement these curriculum requirements:

Table 1.2 D&T programmes of study 1990

Key Stage 1	Key Stage 2	Key Stage 3	Key Stage 4
Pupils should be given opportunities to develop their design and technology capability through:			
3 statements	3 statements	3 statements + 2 bullet points	3 statements
Pupils should be given opportunities to			
3 statements	3 statements	2 statements	1 statement
Design skills			
6 statements	7 statements	12 statements	12 statements
Making skills			
6 statements	7 statements	11 statements	12 statements
Knowledge and understanding			
Mechanisms	Materials and components	Materials and components	Materials and components
Structures	Control structures	5 statements	6 statements
Products and applications		Systems and control	Systems and control
Quality	Products and applications	7 statements	3 statements
Health and safety			
Vocabulary	Quality	Structures	Products and applications
		5 statements	7 statements
7 statements + 2 bullet points.	Health and safety	Products and applications	Quality
		6 statements	5 statements
	Vocabulary	Quality	
		4 statements	Health and safety
			3 statements
		Health and safety	
		3 statements	
	11 statements and 3 bullet points		
Total statements 25	Total statements 31	Total statements 58	Total statements 52

Besides the complexity of the proposals, one of the major problems was the difficulty in interpreting what some of the statements meant. They had been deliberately kept at a level of generality to try and avoid prescription but, even with examples, this meant that the various statements were somewhat abstract or vague.

(McCormick 2002: 37)

Although the terms of teference to the D&T Working Group had contained references to specific materials and components, the long list of content in the programme of study did not (Eggleston 1996). On the one hand, this was helpful, as it gave teachers the opportunity to use whichever materials and teaching methods they felt equipped to use. On the other hand, it added to the confusion and lack of clarity about what was needed and led some teachers to continue much as they had done, re-branding existing projects and work as 'technology'.

Many teachers felt that the new curriculum threatened practical work. Concerns were raised over the increased knowledge and theoretical base that pupils were required to cover, for example in electronics, pneumatics or food science. This was addressed by the NCC in 1991, which stressed that practical work was a core competency within the subject:

> D&T capability empowers people to operate effectively, creatively and confidently in the made world and the Order for Technology assumes that practical engagement by pupils in the processes of D&T is fundamental to an education with this aim.
>
> (NCC 1991: 3)

This uncertainty felt by teachers was exacerbated by the fact that they found it difficult to access guidance documents and support that had been promised; post-introduction meetings at which professionals could discuss and clarify their understanding of the requirements were not readily available (Benson 2000).

Debate continued to rage not only about the content of the subject, and the methods by which pupils were exposed to it, but also about the purpose of the subject. Mulberg (1992) identified that the traditional view of pupils' making was no longer to be accepted, that the subject had to look beyond the manufacture of a product and focus on how both design and technology can respond to a need and improve or enhance individuals, groups or society.

The purpose of D&T was no longer to produce a skilled, manual workforce but to educate more rounded citizens able to draw on a set of skills for both life and work:

> Although it will provide a sound and extremely useful foundation for professional training, Design and Technology Education for 5 to 16 is not intended

to make every child a professional designer or technician any more than mathematics or science are intended as groundings for future professionals in those fields.

(Eggleston 1990: 37)

As the National Curriculum began to establish itself in schools, it brought a growing optimism that design and technology would benefit from an improved status:

One of the most exciting aspects of the National Curriculum is the higher profile given to design and technology; the 1990s could be the time when this subject area and its relationship to our culture receives the wider recognition it deserves. A lot of enthusiasm will be needed to exploit the full potential of the opportunity whilst the initiative remains fresh.

(Norman 1990: 96)

During its first year of implementation of the National Curriculum, the NCC reviewed its impact and received both positive and critical feedback. As a result of this they developed a series of non-statutory guidance documents to help schools with issues such as curriculum structure, the role of subject leaders and developing schemes of work, with some exemplar materials. The idea of producing a national scheme of work (particularly for primary schools) was considered but it was decided that this would be too prescriptive (Benson 2000) and that teachers should be allowed to develop their own schemes and projects. Significantly, one of the objectives of the guidance for technology was to clarify the nature of the subject, something the Working Group had earlier suggested and which later proved to be needed.

National Curriculum developments

In 1991 the NCC publication *Aspects of National Curriculum design & technology* set out to further explain what the 'new' aspects of the subject were and what its educational contributions were:

Firstly, never before has an attempt been made to teach D&T to all children through 11 years of compulsory schooling. To engage the interests and sustain the motivations of all pupils, it would not be enough to simply extend the provision of existing precursors ... the requirement is for something broader and more generally relevant in terms of outcomes, contexts and operation and kinds of knowledge and skills involved ...

Secondly, D&T differs from other subjects in the National Curriculum in that it does not have an established tradition of teaching and learning ... Such traditions are not forged overnight, however, and one of the tasks associated with the implementation of the D&T in the National Curriculum is the creation of a subject culture with which teachers can identify.

> Thirdly, the departure is new in that there has obviously been little research into pupils' understanding and learning in D&T ... it would be invalid to make inference from past performance in, say, CDT and home economics.
>
> (NCC 1991: 2–3)

This statement emphasises the broader approach, links to other subjects and the development of non-traditional projects that could be achieved. Crucially, though, it made the point that continuing to teach CDT and home economics projects was not appropriate; more relevant knowledge and skills needed to be developed. Cross-curricular themes, including economic and industrial understanding, were also to be developed. The NCC was also keen to emphasise the importance of teaching values, ensuring that pupils considered aesthetic, moral, environmental and spiritual issues.

For technology, however, curriculum implementation still faced uphill struggles. Teachers had still not received the support and training needed to change their practice and uncertainty about the nature of the subject, and its place on the school curriculum persisted. There was intense debate and lobbying about the technology curriculum from two particular groups, educationalists and industrialists, particularly the Engineering Council, supported by the Design Council (Wright 2008). The Engineering Council, which had been represented on the Design and Technology Working Group, was 'alarmed' at the way in which technology was evolving and it commissioned research into the teaching of technology (Smithers and Robinson 1992). This report reinforced the teachers' lack of clarity about what was required and concluded that the subject was not clearly defined and 'lacks identity' (Smithers and Robinson 1992: 13).

The NCC had been monitoring the implementation of the National Curriculum and their report, together with others, led to a revision of the technology curriculum in 1992. They reduced the number of attainment targets from four to two, a recommendation from Smithers and Robinson; these were 'designing' and 'making'. However, designing was made up of two strands and making of three, so the streamlining was not as evident as it would first appear. Curriculum content was under the headings:

- construction materials and components
- food
- control systems and energy
- structures
- business and industrial practices.

This is much reduced from the original version.

This curriculum did not last long, however, as in 1993 there was a review of the whole National Curriculum and, such was the discussion over technology, in 1994 the government announced that it was no longer a statutory requirement on the curriculum. Some CDT and home economics teachers breathed a sigh of relief and resumed their earlier practice. Others, unwilling to lose ground and knowing that it would return in one form or another later, continued to teach technology.

The NC review of 1993 became known as the Dearing Report, after its author. As a result of this report, working groups were re-established, this time for design and technology, not merely technology. Recommendations were published in 1994 and implemented in 1995. This new curriculum brought greater clarity to the subject (McCormick 2002; Breckon 2009) and was met with relief and acceptance by teachers. The links to business studies, industry and enterprise and the relationship to other subjects was made clearer, pupils would 'apply skills, knowledge and understanding from the programmes of study of other subjects, where appropriate, including art, mathematics and science' (DfE/WO 1995: 2). Industrial practices were required only at examination level, not in lower secondary schools, which many teachers welcomed and information technology was no longer subsumed within technology but became a curriculum subject in its own right.

The revised content also made it clear that pupils were to be encouraged to design and make products, which had been in the terms of reference but not the original curriculum: 'Pupils should be taught to develop their design & technology capabilities through combining their *Design* and *Making skills with Knowledge and understanding* in order to design and make products' (DfE/WO 1995: 6).

While these revisions were welcomed by many, there was some concern that food became an option rather than a compulsory part of the curriculum, highlighting again the difficult relationship that has always existed. For many, however, the identity of the subject became clearer and its place on the curriculum more assured.

Developments since 2000

Following the five years of stability recommended by the Dearing Report the curriculum remained largely unchanged until September 2000. In 1999 there was a further whole curriculum review and a new D&T curriculum was published (QCA 1999). The subject content had one main heading 'skills, knowledge and understanding' and was set out under headings to reflect the design and make process:

- developing, planning and communicating ideas
- working with tools, equipment, materials and components to produce quality products
- evaluating processes and products
- applying knowledge and understanding.

Food technology teachers were pleased to see that the revised curriculum encouraged the study of food. There was a strengthening of technology within the subject, through CAD/CAM (see Chapter 10) and the requirement for pupils to learn about new technologies and new materials, including smart and modern materials. It further identified that pupils should learn though focused practical tasks and product analysis, as well as design and make activities. The attainment targets were reduced again, to just one 'designing and making'. This brought significant change within the subject and left Kimbell (1999: 3) feeling that the new orders, although offering several benefits, were 'less designerly, less entrepreneurial, less challenging ... and conversely it was more straightforward, more limited and more safe'.

The curriculum was reviewed again in 2007 and all subjects were required to present their content under the same headings, which were:

- Key concepts – for D&T, these were identified as designing and making, cultural understanding, creativity and critical evaluation, although some of these are clearly not concepts.
- Key processes – here the list of processes are those thought to be involved in designing and making.
- Range and content – which specifies the knowledge to be covered, under the headings designing, food, resistant materials and textiles, systems and control.

There was also a list of 'curriculum opportunities' that describe some of the learning experiences pupils should have. The prescribed content was much reduced but 'explanatory notes' were provided alongside the text to make clear what was required. In the content, food was pitted against textiles in that pupils' study should include 'at least one of food or textiles' (QCA 2007: 55), although no rationale was given for this option.

A new UK government, elected in May 2010, began immediate reform of educational provision in England, including a review of the National Curriculum. The outcome of this is still awaited, and the position of D&T is uncertain. The future, therefore, for design and technology in England could be in the hands of schools and teachers – which would open up a world of interesting possibilities.

Conclusions

This brief summary has shown that the introduction and implementation of design and technology on the school curriculum in the UK – and latterly England – has been a bumpy ride and continues to be one. The creation and implementation of design and technology saw several attempts by policymakers to define and position the nature of the subject, and there can be no doubt that they faced a difficult task in creating a new subject. The subjects that contribute to D&T already had a long and mostly successful place on the curriculum and the initial introduction of technology was so badly communicated and managed that much damage was done to those subjects and an opportunity was missed to create an innovative, exciting and relevant curriculum.

The early years of technology were difficult and there seemed a clash between the new design and problem-solving approaches and the traditional skill-based approach. Much of the supporting guidance urged cross-curricular and problem-solving work and discouraged traditional approaches. Despite the 'new horizon' and the opportunity to move the subject forward it was not clear enough to teachers how, why, by what methods and what subject matter they should do this. The subject was in a precarious position, its links to other subjects, particularly science, did not seem strong and it lacked a vision or clear identity.

There was also the issue of food technology within design and technology, this debate constantly keeps resurfacing and has never been properly resolved. (See Chapter 8 for a further discussion of this.)

Various iterations of the curriculum gradually brought greater clarity and as the National Curriculum evolved and developed so did D&T, but it has never fully recovered from the initial mishandling and still finds itself continually having to redefine what it is, explain itself to others and justify its place on the curriculum.

Questions

1   How would you define 'design and technology'?
2   What is the educational value of design and technology?
3   Should we have a National Curriculum that determines what we teach in our individual schools and classrooms?

References

Benson, C. (2000) 'Ensuring successful curriculum development in primary design and technology' in Eggleston, J. (ed.) *Teaching and learning design and technology: a guide to recent research and its application*, London: Continuum.
Breckon, A.M. (2009) 'National Curriculum review 2000', *Journal of Design and Technology Education* 3(2): 101–103.
DES (1981) *The school curriculum 1981*, London: HMSO.
DES/WO (1989) *Proposals of the Secretary of State for Education and Science and the Secretary of State for Wales: design and technology for ages 5 to 16*, London: HMSO.
DfE/WO (1995) *Design and technology in the National Curriculum*, London: HMSO.
Eggleston, J. (1976) *Developments in design education*, London: Open Books.
Eggleston, J. (1990) *Delivering the technology curriculum: six case studies in primary and secondary schools*, Stoke-on-Trent: Trentham Books.
Eggleston, J. (1996) *Understanding design and technology*, Buckingham: Open University Press.
Gillard, D. (2011) *Education in England: a brief history*, www.educationengland.org.uk/history (accessed 29 December 2011).

Harris, M. and Wilson, V. (2003) 'Designs on the curriculum? A review of the literature on the impact of design and technology in schools in England', *Journal of Design and Technology Education* 8(3): 166–171.

Kimbell, R. (1999) 'Coming of age', *Journal of Design and Technology Education* 4(1): 3–4.

Kimbell, R. and Perry, D. (2001) *Design and technology in a knowledge economy*, London: Engineering Council.

Kingsland, J.C. (1969) 'In search of an alternative road' in Baynes, K. (ed.) *Attitudes in design education*, London: Lund Humphries.

McCormick, R. (2002) 'The coming of technology education in England and Wales' in Owen-Jackson, G. (ed.) *Teaching design and technology in secondary schools: a reader*, London: RoutledgeFalmer.

Mulberg, C. (1992) 'Beyond the looking glass: technology myths in education' in Budgett-Meakin, C. (ed.) *Make the future work*, Essex: Longman.

NCC (1991) *Aspects of National Curriculum design and technology*, York: National Curriculum Council.

Norman, E. (ed.) (1990) *Teaching design and technology 5–16*, London: Longman.

Penfold, J. (1988) *Craft, design and technology*, Stoke-on-Trent: Trentham Books.

QCA (1999) *Design and technology: the National Curriculum for England*, London: DfEE/QCA.

QCA (2007) *Design and technology programme of study for Key Stage 3 and attainment target*, London: HMSO.

Smithers, A. and Robinson, P. (1992) *Technology in the National Curriculum: getting it right*, London: Engineering Council.

Wright, R. (2008) 'The 1992 struggle for design and technology', *Design and Technology Education: an International Journal* 13(1): 29–39.

Chapter 2
Developments in teaching design and technology

Daniel Wakefield

Introduction

This chapter develops from Chapter 1 and debates the change of practice in design and technology teaching in the UK since the introduction of the National Curriculum (NC). There have been a number of projects and initiatives created to develop the subject and each has influenced to some extent how D&T is taught and how pupils engage with it.

In considering developments in the teaching of D&T, it must be remembered that there have been significant developments, particularly over the last 10 years, in whole-school approaches to teaching and learning that have had their impact on all subjects. These are not considered here but have had an influence on the general approach to D&T.

Chapter 1 discussed the development of the National Curriculum for design and technology, which created debate around the understanding and position of the subject. It showed that, for some teachers, there has been a sense of confusion or lack of clarity around the purpose and aims of D&T, what pupils should be taught and the approach to be taken. In this chapter, I will consider how some of those issues affected the practice of classroom teachers and how the teaching of the subject has changed, or not, since the introduction of the National Curriculum.

Design and technology in the primary school

The National Curriculum created a huge amount of work for primary teachers, who were required to plan, prepare and teach 10 subject areas in which the content was now prescribed by government. They understandably focused on the core curriculum of English, mathematics and science and, as a result, many of the foundation subjects, including design and technology, suffered (Jarvis 1993).

There was also some concern about the discrete nature of the subject with Bentley, Campbell and Sullivan (1990) suggesting that design and technology would have been better conceptualised and implemented as a cross-curricular subject. Other teachers, mainly female, felt uneasy about what they perceived as craft, design and technology (CDT) being introduced into the primary curriculum as they perceived it as a technical subject requiring a high degree of skill (Smits 1990), which they felt they lacked (Makiya and Rogers 1992). At that time, few primary teachers would have experienced D&T either in their own schooling or their teacher preparation programme (Johnsey 1998). There were also concerns about the lack of training and resources required to teach this new subject (Bentley, Campbell and Sullivan 1990). Other primary teachers, however, welcomed it for the contribution it made to pupils' learning, developing skills and attitudes not previously covered in the primary curriculum.

As the D&T curriculum developed in primary schools, there were some criticisms that pupils, although building models of castles, making purses or making sandwiches, were not really engaging in design and technology activities. Pupils were able to see the relevance of the work and understand the significance of the outcome, but they lacked the design work and had limited understanding about how and why equipment, materials or processes were used (Jarvis 1993). This was not the fault of the teachers. In primary schools, as in secondaries, teachers felt unprepared and unskilled, they did not understand the subject or its requirements and had to make the best they could with the limited information and support they had.

Chapter 4 discusses design and technology in primary schools in more detail.

Teaching design and technology

There have been many influences that have resulted in changing practice within design and technology classrooms. These influences include:

- changes in the National Curriculum (see Chapter 1)
- government guidance on the National Curriculum, including national strategies
- subject-specific initiatives such as Nuffield D&T Project, Licence to Cook, the Technology Enhancement Project and Digital D&T Programme.

Despite or perhaps because of all these influences there are huge variations between schools across the country in design and technology education. I shall, first, outline the influences listed and then consider how these have impacted on the teaching of D&T.

National Curriculum

As Chapter 1 describes, the introduction of technology on the National Curriculum created mostly a sense of uncertainty among most teachers – what were they supposed to be teaching and how were they supposed to be teaching it? It is hard to convey how bereft teachers felt, they could no longer teach the subject they had trained to teach and which they had been teaching successfully for years. Curriculum documents were not helpful in providing guidance as to what knowledge and skills pupils should be taught and many teachers felt that they had been left to work it out for themselves.

Unfortunately, this uncertainty about what was required and the perception that D&T was 'new' led to the loss of good practice that had been developed over decades in the teaching of CDT and home economics. This was replaced with new, untried approaches and work requiring less skill from pupils, particularly in making (Mason and Houghton 2002).

With the lack of support and guidance from government, many schools turned to the National Design and Technology Education Foundation (NDTEF). NDTEF provided professional development for teachers and promoted a cross-subject theme approach to teaching, with many departments using themes such as survival, transport or the environment to link the different areas now within D&T. This was not always fully understood, however, and one ex-home economics teacher proudly told a colleague that she had not switched on a cooker for weeks (personal communication).

In 1995, with a revised curriculum that separated information technology and positioned D&T as a more discrete subject (McCormick 2002), more traditional forms of teaching reappeared. Schools started to teach through separate focus areas such as food, systems and control, resistant materials and textiles. The subject became timetabled in regular slots and schools shaped the curriculum not only to accommodate curriculum requirements but also around the skills and expertise of the teachers and the resources available.

This, however, created its own drawbacks. Pupils often experienced the different areas within D&T through a 'carousel approach' in which, over a school year, they spent a few weeks in each area so that by the end of the year they had covered all the areas.

Unfortunately for the pupils, there was often little communication between the teachers in the different areas and so pupils found themselves following the same 'design and make' process over and over again, responding to a design brief and working through a number of stages: specification, initial ideas and final design before moving onto the manufacturing stage of the project then concluding with a written evaluation. The only difference each time was the materials used. Teachers took no account of pupils' previous experiences in D&T so there was little progression in their learning and they often became bored and demotivated. It also led to pupils perceiving D&T as four or five separate subjects.

The carousel option also created another difficulty in that pupils had only a few weeks in each area and any interruption, such as a bank holiday or school trip, reduced the teaching time even further. This meant that boxes of unfinished projects sat in teachers' store rooms, leaving both pupils and teachers feeling frustrated.

Food technology teachers also were critical of the 'design and make' pedagogical approach advocated. In order to be able to assess pupils in the same way in food as they were assessed in other areas of D&T, food teachers were expected to ask pupils to 'design' a food product. This was not an approach that they were familiar with, and it led to pupils being asked to 'design' pizzas, sandwiches, salads or cake toppings. Pupils spent time drawing food products before making them, a practice which surely cannot be supported, but which food teachers felt was a necessary part of the design and make process that all areas of D&T were expected to follow.

Since the introduction of the first National Curriculum there have been numerous initiatives developed to support teaching and learning in design and technology. There is not enough space in this chapter to consider them all, so those discussed here are only a representative sample.

National strategies

The national strategies have had a major impact on the teaching of D&T in schools in England. They were established by the DfES in 1998, initially focusing on literacy and numeracy in primary school but later expanding. The strategies were designed to raise standards in pupils' attainment through supporting and developing the quality of teaching and learning within school.

In 2002, building on work in the primary sector, the Key Stage 3 (lower secondary) strategy was launched. There were five strands: English, mathematics, science, information technology and the foundation subjects, starting with D&T. The four key aims of the strategy for D&T were to:

- increase expectations and challenges in teaching
- define progression through the yearly objectives in key skills and concepts to facilitate planning, teaching and assessment
- improve the teaching of design
- improve the quality of teaching in design and technology by providing examples of effective teaching approaches, drawn from the training materials for foundation subjects, that include *questioning, modelling* and *structuring lessons*. (DfES 2004: 11)

Learning lessons from the introduction of the National Curriculum, the Department for Education funded a National Strategies Unit, which provided training programmes across the country to help teachers implement the strategies (which later expanded even further). The framework provided teachers with detailed guidance on how to develop or improve their teaching; examples were given of lesson planning, short-, medium- and long-term department planning, curriculum development and curriculum tracking. What proved particularly successful with D&T teachers were the examples of worksheets and activities for starting lessons and activities for developing or improving design work. Although, overall, reaction to the framework was mixed.

The National Strategies Unit was closed in 2009 and its website in 2011, but the legacy of the framework can still be seen in many schools and the national strategies folder will be found (somewhere) in most departments.

Subject-specific initiatives

Nuffield D&T Project

The D&T Project was established in 1990 to support teachers in implementing the National Curriculum. The philosophy behind the Project was that pupils should be able to 'design what they could make and make what they could design' (Nuffield D&T Project online, www.nuffieldfoundation.org/nuffield-design-technology (accessed 13 February 2012)). Nuffield produced a wide range of resources for both primary and secondary D&T and supported these with training and support for teachers. These resources made a major contribution to the development of teachers' practice, both primary and secondary, especially through the use of 'resource tasks' and 'capability tasks' to support pupils in developing knowledge and skills which they then utilised in design and make activities. This pedagogy is still evident in many classrooms.

Licence to Cook

The Licence to Cook initiative ran from 2008 until 2011. It provided a statutory entitlement for all pupils to learn how to cook and, in theory, to learn about nutrition, food safety and consumer issues, although in practice these aspects were often overlooked. The government funded training and resources to support the introduction of this initiative and many food teachers responded positively. However, concerns have been expressed about the damage that Licence to Cook has caused to food technology (Rutland 2008, 2010). Rutland notes that, although the initiative raised the profile of food in schools and provided additional funding, many teachers taught Licence to Cook in food technology lessons, so that pupils learnt the practical skills but missed out on the 'knowledge, understanding and experience of the wider issues of food technology' (2010: 9).

As discussed in Chapter 1, food has had an uneasy alliance with some of the other technology areas. Like CDT, its history was as a practical subject supporting the development of life skills and providing skills for those interested in employment in a related industry. With the advent of food technology, many home economics teachers felt that the essence of the subject had been lost and embraced the Licence to Cook initiative with its emphasis on practical cooking. Food in the school curriculum seems to be in a state of uncertainty (this is discussed further in Chapter 8).

Technology Enhancement Project

The Technology Enhancement Project (TEP) was set up in 1990 'to enhance technical education and training in UK schools through the production and distribution of resources to support teaching and learning' (Gatsby, www.gatsby.org.uk/en/Education/Projects/Technology-Enhancement-Programme.aspx

(accessed 13 February 2012)). Supported by Middlesex University and the Gatsby Charitable Foundation, TEP has produced a number of publications and a vast range of resources and training programmes for teachers, all designed to support the teaching and learning of D&T.

The resources are usually low cost and accessible to teachers, they build on teachers' existing knowledge and skills and help them develop a more informed approach to teaching D&T. As the D&T curriculum has developed, and as new materials and technologies have emerged, TEP has kept pace with these so that their resources also help teachers to keep up to date with the projects they offer pupils.

Although it has the general name of 'technology', TEP focuses on supporting the teaching in the resistant materials and electronics areas and more lately engineering, although they do produce some resources for textiles technology. Teachers have particularly welcomed their packs of smart materials, PIC resources and project packs.

Digital D&T Programme

The Digital D&T Programme has grown out of two previous initiatives, CAD/CM in Schools and the Electronics in Schools Strategy. It is funded by the DfE to provide professional development for teachers.

Computer-aided design/computer-aided manufacture (CAD/CAM) was introduced to the D&T curriculum in 1999 and the CAD/CAM in School Initiative was launched in response to this. The initiative was funded by the DfEE and provided free 3D design software to schools, along with training for teachers in its use and implementation. A number of pilot schools and hub centres were established to provide software training and resources to enable teachers to gain a greater understanding of CAD. The initial findings from the pilot schools were that many benefited from the initiative, staff became competent users of the software and embedded CAD/CAM into their curriculum. Breckon (2001) found that this initiative had a more immediate impact than many others.

Although the initiative may be considered prescriptive, in that it controls the software used in schools and, to some extent, how it is used, the underlying philosophy gives a different picture:

> [T]he focus of the project is more on the children than on the technology. Although the theme of the project is CAD, it is used to develop other aspects of a child's character and learning abilities. Design in schools, more than any other part of the syllabus, provides children with something that has an outcome. Children are not so good at carrying things conceptually, and here they are able develop ideas through to a pleasing and successful conclusion.
>
> (Breckon 2002)

The Electronics in Schools Strategy was established in response to the growing availability of electronics for use in schools and the lack of teachers' qualified to teach electronics. The Strategy set up regional centres and provided free training for teachers and funding for resources, providing that the teachers continued to teach electronics in their schools.

The Digital D&T Programme incorporates both of these initiatives and continues to support and promote the teaching of electronics and CAD/CAM in schools.

How has design and technology changed?

Traditionally, practice in CDT and home economics was for pupils to simply produce copies of practical work produced by the teacher, particularly at lower secondary level. The teacher demonstrated a process and each pupil would copy and attempt to produce the same product. Little attention was paid to why the item was being made, why it had been designed in a particular way or why certain materials or processes were employed. This approach often led to difficulties at examination level, when pupils struggled to work independently.

Since the shaky start in 1990, there has been a change in the philosophy towards D&T, from it being a subject dominated by practical skills to one that, at its best, challenges pupils and engages them in decision making, problem solving, analysing, justifying, evaluating as well as developing practical skills. As Chapter 1 discussed, historically the subject is rooted in the traditional 'making' mould and, in some places, that perception still remains. However, the influences outlined earlier have contributed to attempts to change that perception.

The National Strategy Framework for D&T impacted on teachers' practice in two ways; it introduced the 'three-part lesson' and it suggested activities for improving pupils' designing skills. The three-part lesson comprised a starter activity, main learning episodes and a plenary (although this has since been adapted). Starter activities are designed to engage pupils in the lesson and bring their attention to the topic, plenaries to summarise and evaluate what has been learnt. Initially, D&T teachers claimed that lessons were too short for these, particularly practical lessons where pupils needed to get in and get on and needed time at the end of the lesson for clearing up. However, it has been shown that starter activities can be effective in refocusing pupils' thinking on the work to be done and so allowing them to make better progress. D&T teachers are therefore, along with other subject staff, being encouraged to use starters and plenaries in all lessons.

The framework materials for supporting the teaching of design skills included a number of activities, for example 'six thinking hats' and 'walk on the wild side', to introduce teachers to different and more interesting ways of approaching the teaching of design. These were embraced by many schools and are still much in evidence.

The Nuffield D&T Project's main influence was the idea of teaching pupils knowledge and skills through resource tasks and capability tasks, which they then utilised in their design and make activities. Most teachers welcomed this approach as it allowed for the teaching of knowledge and skills, which they felt had been overlooked in some of the early NC programmes. The Nuffield resources also helped teachers to see how they could work with colleagues to develop more coherence in the subject. By identifying more clearly some of the common aspects of work across the different areas within D&T, teachers could see that pupils' learning about how to write a specification, materials list or evaluation in one area could be developed in another area. As Rutland (2002: 62) noted:

> Much progress has been made since the introduction of Technology into the National Curriculum in 1990. There is now a better understanding of the common, unifying elements of the subject and an increased appreciation of the differences across the different contributing areas. However, the outcomes for the pupils ... are fundamentally the same across all the areas of Design and Technology.

Licence to Cook, unfortunately, has not had the same positive effect. The focus of the initiative was on pupils learning practical skills, almost by rote or the apprenticeship model. Teachers who embraced this approach, and there were many, have not helped pupils' learning in food technology. It is possible in food technology to teach pupils cooking skills while also teaching them the principles of food technology. For example, a lesson on making scones could also get pupils to cook one of their scones using a different method from baking. Pupils could then compare scones made using the traditional method and those cooked using steam, boiled in water, shallow fried or cooked in a microwave oven. This could, then, lead into a discussion of the principles of the different cooking methods and why each is suitable for different dishes.

The Technology Enhancement Project and Digital D&T have, in contrast, helped to create a vision for design and technology around modern materials and processes. Both initiatives identified the possibilities for new and innovative projects and created high-quality resources that are financially accessible for schools (Breckon 2000). Through using these resources and project ideas teachers have updated their own knowledge and pedagogy to ensure that pupils have a contemporary and challenging learning experience in D&T.

These projects have each contributed something to the development of the teaching of D&T, from the provision of resources to the ways in which teachers introduce knowledge and skills to pupils. Design and technology is a subject in which the content will continue to evolve as new materials, processes and equipment emerge and teachers have to be prepared to continually evolve their practice. The benefits of teachers continuing to develop their pedagogy is that pupils are more engaged in the subject than ever before, teachers challenge pupils to think more creatively, become more independent and take risks, all of which is vital to the continuing success and relevance of the subject.

Conclusion

The teaching of D&T in England and Wales has developed considerably since the introduction of the NC. If we compare practice in the early 1980s with that of today there is a clear change in philosophy, attitude and approach to the subject. The projects discussed here, and others, have contributed to this. In many schools across the country, pupils are now more engaged with their work and are developing their knowledge within D&T. Teachers use different techniques and resources allowing pupils to gain understanding and experiences which can then be used to develop project work.

The development of D&T pedagogy has not been an overnight process; it has taken many years of development, as well as support and encouragement from outside agencies. It could be argued, however, that the D&T community now needs to take more ownership of curriculum and pedagogy developments, there needs to be more research and more teachers developing practice through taking risks. The National Curriculum created opportunities for teachers to develop new, exciting and innovative ways of teaching. The proposals for D&T (DfE 2011), if adopted, could continue this trend and D&T pedagogy would embrace change and challenge. If not, who knows what the future holds?

Questions

1. How do teachers reflect their understanding of D&T through their teaching?
2. How much of teachers' practice should stay the same and how much should it change?
3. How much should teachers rely on outside agencies for introducing change or how much should they take responsibility for it themselves?

References

Bentley, M., Campbell, J. and Sullivan, M. (1990) 'Primary design and technology' in Bentley, M., Campbell, J., Lewis, A. and Sullivan, M. (eds) *Primary design and technology in practice*, Essex: Longman.

Breckon, A. (2000) 'DfEE/DATA CAD/CAM in Schools Initiative – a success story so far', *Journal of Design and Technology Education* 5(2): 153–157.

Breckon, A. (2001) 'DfEE/DATA CAD/CAM in Schools Initiative: the designing and making revolution in design and technology education', *Journal of Design and Technology Education* 6(2): 161–166.

Breckon, A. (2002) 'CAD user', *Mechanical Magazine* 15(6) www.caduser.com/reviews/reviews.asp?a_id=142 (accessed 27 September 2012).

DfE (2011) *The framework for the National Curriculum: a report by the Expert Panel for the National Curriculum review*, London: HMSO.

DfES (2004) *Design and technology: framework and training materials Key Stage 3 National Strategy*, London: HMSO.

Jarvis, T. (1993) *Teaching design & technology in the primary school*, London: Routledge.
Johnsey, R. (1998) *Exploring primary design and technology*, London: Cassell.
McCormick, R. (2002) 'The coming of technology education in England and Wales' in Owen-Jackson, G. (ed.) *Teaching design and technology in secondary schools: a reader*, London: RoutledgeFalmer.
Makiya, H. and Rogers, M. (1992) *Design and technology in the primary school*, London: Routledge.
Mason, R. and Houghton, N. (2002) 'The educational value of making' in Sayers, S., Morley, J. and Barnes, B. *Issues in design and technology teaching*, London: RoutledgeFalmer.
Rutland, M. (2002) 'Links across design and technology' in Owen-Jackson, G. (ed.) *Teaching design and technology in secondary schools: a reader*, London: RoutledgeFalmer.
Rutland, M. (2008) 'Licence to Cook: the death knell for food technology?' in Norman, E.W.L. and Spendlove, D. (eds) Design and Technology Association International Research Conference, Wellesbourne: Design and Technology Association.
Rutland, M. (2010) *Food technology in secondary school in England: its place in the education of a technologically advanced nation*. Paper presented at Design and Technology Association Education and International Research Conference, Keele University, July 2010.
Smits, A. (1990) 'Primary school technology: where is it going?' in Bentley, M., Campbell, J., Lewis, A. and Sullivan, M. (eds) *Primary design and technology in practice*, Essex: Longman.

Chapter 3

International perspectives on technology education

Frank Banks and P. John Williams

Introduction

If you are a design and technology (D&T) teacher or student teacher in the United Kingdom, we guess you know a lot about the D&T curriculum that you teach. As teachers, we get to know the details of the syllabus and examination specifications that we use, but rarely question what we teach – or why. Here we look at the technology curriculum of some selected countries and ponder why there are differences. We hope that this will contribute to the debate about the nature and purpose of D&T and that you will reflect on *why* the schemes of work that you are currently teaching are as they are – and what could or should be different? Are there lessons to learn from other countries?

In looking elsewhere for curriculum ideas you will not be alone, Pavlova (2006: 20) notes: 'People involved in the development of technology education were looking around the world for ideas.' Those ideas assumed there was a common goal for technology education that was, at least at the broad level, vocational in its aim to create a flexible and adaptable workforce. Pavlova summarised:

- in the UK, the former Secretary of State for Education, Kenneth Baker, announced that technology as a subject was considered to be 'of great significance for the economic well-being of this country'
- in Australia, a statement on technology for schools explained 'Technology programs prepare students for living and working in an increasingly technological world and equip them for innovative and productive activity'
- in the USA, it was announced that technology education was 'vital to human welfare and economic prosperity'.

Comparative research has shown that the development of technology curricula across the world has been slow and implementation restricted, even when the new subject is 'compulsory'. For example, Ginner (2007) noted similarities between the technology curricula of Sweden and New Zealand but counselled that just because it is prescribed, does not mean it is actually taught:

It has been and still is a challenge to implement Technology in the Swedish schools. ... given the fact that it is a compulsory subject the number of schools neglecting the area or doing too little is surprisingly high. . . . The coming years will be decisive – at least in Sweden. It could finally take its place and be accepted as a natural and vital part of the curriculum or gradually evaporate and/or merge into science.

(Ginner 2007: 2)

Ginner is clear that just specifying that technology should be taught is not enough to ensure implementation or that it is taught as the curriculum designers intend.

Each country builds on its history of technical education and develops an approach within its own context to suit the perceived needs of society and the individual. But first, why teach technology and how might it be taught?

Why teach technology?

McCormick (1993) suggested four justifications for why technology should be part of the curriculum:

1 The personal development opportunities it provides for students, for example practice in the solution of real problems and the associated thought processes and the multidisciplinary approach to knowledge and information essential to technology education. This is stated as a rationale for technology education in Australia.
2 Education for the technological culture in which we live, to enable students to become informed decision makers and responsible users of technology, not so much for their own sake but for the benefit of society, a significant rationale for technology education in South Africa.
3 The vocational dimension of technology education is a rationale that comes and goes with the passage of time and tends to correlate with periods of national economic depression when policymakers and industrialists turn to education as part of the solution (Williams 1993).
4 Technology education as education for production was a strong rationale in many Marxist-driven economies. With the collapse of the Soviet Union, this rationale is less common, but was a driving force in eastern European countries such as Hungary, Czechoslovakia, the former German Democratic Republic and southern African countries such as Zimbabwe and Mozambique.

In the 2007 version of the national D&T curriculum for 11 to 14-year-olds in England – which is currently being taught in schools – although there is mention of innovation and consideration of 'industrial issues', in contrast to the original aims of the National Curriculum there is little that is explicit about 'economic well-being' or 'economic prosperity'. This curriculum built on a succession of curriculum documents and consultations in England (see Chapter 1) and, unlike some past curricula, was well received by teachers. However, despite its quiescent reception, the statement is as political as the original, both in what it says about D&T education and what it does *not* say. It highlights 'creative thinking', 'current technologies', 'problem solving' and 'design' (a form of problem solving) and suggests that 'pupils develop confidence in using practical skills . . . as individuals and members of a team' (QCA 2007: 51). These terms are loaded. For example, there has been much critical discussion as to whether pupils can truly be 'creative' when, under a rigid assessment process, teachers 'play it safe' and direct pupils' design and making closely to ensure they gain high marks. Similarly, whether, in reality, teamwork is compatible with the making of individual products. There is also some disquiet about the balance between traditional craft skills and current technologies. It will be interesting to see if these concepts are retained in the next curriculum review.

Some also criticise what is *missing* from such a definition of technology education. Keirl (2007) and Petrina (1998, 2007) believe that technology education should consider the politics and values involved in making choices of products and materials. They maintain that consideration of 'who wins, who loses' in the manufacture of, for example, footballs in China or fashion garments in India should be part of every pupil's education. Similarly, Elshof (2005, 2006) and Hill (2006) make a case for environmental sustainability as being a more important driver of 'technology education for all' than design-and-make assignments.

How should technology be taught?

De Vries (1994) proposed a number of categories of approaches to technology education, which have been cited in many contexts, including Layton (1993) and Black (1996):

1   A tradition in many countries has been the teaching of craft skills, often through the construction of set projects and the repetitive practice of relevant skills. This is a part of the basis of the Swedish tradition of *Sloyd*, which has influenced the development of technical education in many countries and is still one of the approaches utilised in Sweden and in many parts of the UK.
2   In some instances, technology education is organised along the lines of mass production, often in a business-like framework. Relevant skills relate to the use of jigs and fixtures, a production line sequence of activity, control and organisation. This is utilised in some eastern European countries and to a lesser extent in a manufacturing technology context in the USA.
3   Although generally rejected as appropriate for technology, it may be organised as 'applied science', where technology is used in the teaching and learning of science, as in Denmark. Sometimes technology gets played down where it is integrated with science and is not dealt with as valid in itself, for

example in the subject 'science and technology' in Israel and the emphasis of teaching science, technology, engineering and mathematics (STEM) as a combined approach in the USA and in England (see Chapter 11).

4   A focus on technology as exclusively high or modern technology, which is futuristic and emphasises information technology. France, for example, and some of the learning modules in the USA and the 'current technologies' emphasis in England are inclined toward this approach.
5   Design, while a methodology of technology, may also be its organisational focus. In this case, specific content is not so important but rather there is an emphasis on the *process* through which students proceed in designing solutions to problems. Both the UK National Curriculum and a number of approaches in Australia have been criticised for this approach and proposals for South Africa are inclined in this direction.
6   Technology may be structured as a series of problems to be solved, requiring information that is multidisciplinary in nature. This is a common approach in parts of the USA.
7   The organisation of content around the achievement of competencies is becoming a more common approach, evidenced for example in the 'attainment targets' of the Netherlands and the UK, the 'competencies' in Australia and 'performance targets' in Sweden.

We have selected a range of countries for you to consider in an attempt to represent some of the diversity of technology education throughout the world. We start with the United Kingdom as many teachers in the UK are surprised by the different approaches to technology education that have developed in a relatively short period of time across the four nation states. As you look critically at the examples given in the following, think of which of McComick's four 'justifications' might predominate and whether the shortcomings suggested ring true.

United Kingdom

The British government of the late 1980s introduced a prescribed curriculum into state schools in England, Wales and Northern Ireland, detailed in Chapter 1. However, the nature of technology is not the same across the UK as, over the years, national parliaments and assemblies have become increasingly more responsible for education policy and practice and national variations in 'technology' have become more marked.

England

As Chapter 1 describes, the curriculum has undergone repeated revisions and is awaiting further revision. The main thrust of D&T in England is using the developed understanding of materials, components and systems for the development of products. The current curriculum (2007) describes D&T in terms of four 'key concepts' (or processes):

- designing and making
- cultural understanding
- creativity
- critical evaluation.

The National Curriculum in England is described in detail in Chapter 1.

Northern Ireland

In technology and design, as it is named in Northern Ireland, the aim is to enable pupils to become confident and responsible in solving real-life problems, striving for creative solutions, independent learning, product excellence and social consciousness. There are nine curriculum areas and with technology and design as part of the science and technology area there is a strong 'applied science' thrust; students should acquire, develop and apply:

- scientific knowledge and understanding
- a range of intellectual skills
- a range of physical skills
- a range of communication skills
- an understating of science and technology's effects on community, economy and the environment.

These are achieved mainly through the designing and making of products in resistant materials, 'product design' or 'systems and control' where the emphasis on electronics is sophisticated compared to other areas of the UK.

Scotland

Following consultation, guidelines for a new 'Curriculum for Excellence' were implemented in 2011. The curriculum includes an area called 'technologies' which aligns with creative, practical and work-related activities that can be applied around six 'organisers':

- technological developments in society
- ICT to enhance learning
- business
- computing science
- food and textiles
- craft, design, engineering and graphics.

'Curriculum for Excellence' lacks the prescription of the curriculum in other parts of the UK and schools are encouraged to design their curriculum to suit local needs.

As in many countries, what is learnt in 'technologies' is largely determined by the examinations system. There is no examination with the word 'technology' in the title, although courses in craft and design (C&D), graphic communication (GC) and technological studies (TS) are available. Technological studies is very similar to the content of 'Design and technology: systems and control' in England and Wales, and technology and design in Northern Ireland, and is by far the least popular of the three. Increasing in popularity are courses simply called 'Practical skills' available in woodworking or metalworking.

Wales

The design and technology curriculum in Wales specifies separately 'designing' and 'making', including the use of hand tools and CAD/CAM machines. In spirit, the curriculum is like that of England, but it suggests the range of contexts to which pupils should be exposed and in particular:

> Learners aged 7–14 should be given opportunities to develop and apply knowledge and understanding of the cultural, economic, environmental, historical and linguistic characteristics of Wales. Learners aged 14–19 should have opportunities for active engagement in understanding the political, social, economic and cultural aspects of Wales as part of the world as a whole.
> (WAG 2008: 8)

United States of America

The educational system in the USA is decentralised, with each state responsible for its own education, although the federal government provides some general control through funding guidelines. Some states pursue curriculum development and implementation at state level, while others give broad guidelines with the actual curriculum work done by local school districts. This results in great variety.

This diversity in the USA makes it difficult to generalise, but it is probably correct to say that the majority of school programmes have a strong focus on skills development and few focus on design or the processes of technology. Historically developments in the USA have been criticised (Todd 1991) because of their insular approach to technology curriculum developments and this is still true today.

The professional association, International Technology and Engineering Education Association (ITEEA) was previously the International Technology Education Association (ITEA). The addition of engineering was a disputed move, but technology education in the USA has struggled for many years (as in many other countries) with public misconceptions about the nature of technology education, most commonly confusion with information technology and computing. The move to engineering was intended to overcome this as the general population has a clearer conception of engineering than it does of technology.

The other rationale for including engineering is the science, technology, engineering and mathematics (STEM) movement. This is a politically driven attempt to address the declining number of students studying science and engineering at university and the presumed consequential decline in economic activity. Exactly what STEM means in terms of the school curriculum is still unclear, but it has an attractive political ring to it (see Chapter 11 for a discussion of STEM).

The engineering professional association is, of course, interested in promoting the role of engineering in schools and as there is no discrete engineering subject, it looks to technology as a way to promote pathways into university engineering courses. The irony of this situation is that university engineering departments are quite conservative in their prerequisites and often demand mathematics and science rather than a technology or engineering subject.

Developments in the USA are often the result of deliberations of higher education and the academic community rather than grassroots developments. Consequently, the trend in schools, as a result of engineering and the STEM movement, is to change the name of the department to engineering or engineering and technology, but whether the nature or the content of the teaching also changes will be variable from school to school.

In very few areas in the USA does technology constitute part of the core curriculum, which means it has to compete for students with other electives. Because of the relative expense of technology programmes, many have been eliminated from schools when student numbers have dropped.

China

China has a proud history of significant technological inventions, from paper and gunpowder to satellites, nuclear energy, superconductors, high-energy accelerators, advanced computers and robots. In China, however, these developments are not seen as being related to general technology education.

China has a long involvement in vocational and technical education, but not technology education as general education. A curriculum area titled 'Integrated curriculum of practical activity' has been part of the curriculum since the early 1900s. This included information technology, researched-based learning, community service and labour-technical education. Information technology has developed into a subject in its own right, as has labour-technical education. These subjects have now been overtaken somewhat by the development of technology, which was developed from the education curriculum reform that began in 1999 and is currently being trialled in five provinces. In 2001 technology was confirmed as one of the eight learning areas of the compulsory curriculum and by 2003, experimental technology curriculum standards were issued. The trial implementation of the curriculum began in four provinces in 2004, with a population of about 35 million students which involved extensive research including the school context, the social expectations of student development and the psychology of high school students.

The technology curriculum is designed to develop students' practical competence in exchanging and processing information, as well as applying technological principles through design, with the goal of fostering students' initiative, creativity and life-planning abilities.

The high school technology curriculum is organised into compulsory and elective subjects. The compulsory subjects are designed to guarantee a student's basic literacy. The elective subjects allow the students to develop their individual interests and can enhance the adaptability and flexibility of the curriculum when implemented in different regions of China. The two main areas of the curriculum are information technology and general technology.

The information technology curriculum consists of six modules, one compulsory and five elective. The compulsory module, information technology foundation, is to raise students' information technology literacy. The five elective modules are algorithms and programming, the application of multimedia technology, the application of network technology, data management and artificial intelligence.

The general technology curriculum includes nine modules, two compulsory and seven elective. The compulsory part includes technology & design 1 and technology & design 2; the basic content of these is technical design. The seven elective modules offer students the chance to choose according to their interests and are electronic control technology, architecture and architectural design, making simple robots, modern agriculture technology, home economics & life technology, garments and garments design and automobile drive and maintenance. Electronic control technology and home economics & life technology are the most common electives offered, often prior to the offering of other elective modules. Rural high schools tend to offer modern agricultural technology as an initial elective.

The new technology curriculum in China has the following characteristics:

1. It is the first time in China that technology education has been established as a learning field, which is an historic breakthrough in curriculum development.
2. With the goal of improving students' technology literacy, technology education has evolved from traditional skill training into a more contemporary form of technology education.
3. Information technology is strengthened, which manifests the transition from computer education to information technology education.
4. Technical design is the core content, which provides opportunities to cultivate student's initiative, creativity and practical competence.
5. The evaluation system is understood by teachers and a system of technological certification is established.

However, the enormity of implementing a new technology curriculum, in a country with over 300 million students, should not be underestimated.

Sweden

Similar to Finland, Sweden has a traditional technical education history: a vocationally oriented craft technology for boys and a home economics-type subject for girls. Since the 1980s politicians have been concerned about the importance of technology in society and the need for education to prepare students adequately, and in 1994 it was made a core subject for compulsory schooling (6 to 16 years).

In a 1993 proposal for changes in technology, it was described as comprising technological components (tools, machines, systems, etc.) and technological skills and knowledge. Knowledge and experience areas were social science, science knowledge and practical knowledge. Technology was therefore neither applied science nor pure practical skills. The emphasis was that practical skills are a type of knowledge, different from theoretical knowledge, but still a type of knowledge and not just hand work.

The brief to the curriculum developers prevented the specification of content; the curriculum was to be based on 'performance steering by targets' and teachers and pupils were to decide what content to use to achieve the targets.

Some elements of the proposal that were to provide a structure for teachers include:

- links between components, tools or machines and systems
- it must be practical
- operative functions of technology are important (e.g. to transform, to store, to control)
- the effects of technology on society and the environment
- the history of technology.

The objectives to be reached by grade 5 (aged 11), pupils to be able to:

- describe, in some areas of technology they are familiar with, important aspects of the development and importance of technology for nature, society and the individual
- use common devices and technical aids and describe their functions
- with assistance, plan and build simple constructions.

The objectives to be reached by grade 9 (aged 16), pupils to be able to:

- describe important factors in technological development, both in the past and present, and give some of the possible driving forces behind this
- analyse the advantages and disadvantages of the impact of technology on nature, society and the living conditions of individuals

- build a technical construction using their own sketches, drawings or similar support and describe how the construction is built up and operates
- identify, investigate and, in their own words, explain some technical systems by describing the functions of the components forming it and their relationships. (CETIS 2011)

South Africa

Although apartheid ended in 1994, the education system is still changing in South Africa. Education has traditionally reflected the pre-1994 system of government with separate departments for each racial group, overseen by the Department of National Education, with patterns of demographics, subjects studied, and the proportions of students proceeding onto higher levels varying with only 35 per cent of blacks compared with 98 per cent of whites passing matriculation. Expenditure on black education was about one-quarter that spent on white education, with very poor facilities for blacks. It is proving to be a massive task to equalize the educational system with over 12.3 million students, 386,000 teachers and 48,000 schools.

The first national curriculum was implemented in 2006 and consisted of eight compulsory learning areas, one of which was technology, based largely on the English D&T model. However, there were no technology teachers and so the first teachers were co-opted from other areas until others were able to rapidly complete short courses and re-qualify as technology teachers. Technology teacher training is now embedded in university programmes, but there is still a drastic shortage of technology teachers.

The reasons for including technology in the curriculum include enhancing the opportunities of the disadvantaged, the technological nature of society, national economic problems, possibilities for personal development in the higher cognitive skills and creative thinking and problem solving. However, the degree of implementation in schools varies remarkably, from independent schools whose technology facilities compare favourably to any in the world, to schools that have no electricity or water and for whom, consequently, technology as a subject is not a priority.

The curriculum review in 2009 resulted in some changes for technology. In the primary years, technology was combined with science into one learning area. One impact of this is that those students progressing onto the middle years will be less well grounded in the fundamentals of technology. For the middle years, technology remains a discrete learning area with three broad outcomes:

- processes and skills (design process and ICT)
- knowledge and understanding (structures, processing and systems and control)
- society and the environment (indigenous technology, impacts and biases).

At the post compulsory level (15 to 17-year-olds), there is no general technology subject but the technology areas are: electrical technology, mechanical technology, civil technology, engineering graphics and design and computer applications. However, the vast majority of schools in the country will not have either the physical or human resource capacity to offer any of these subjects.

The South African technology curriculum is unique in that it includes a specific outcome focused on indigenous technology. The government has attempted to support schools' and teachers' development in this area for a number of reasons. Most of the technology curriculum reflects 'western' views about technology and a focus on indigenous technologies is an attempt to rectify that. In addition, in a context of a large number of extremely resource-poor schools, teaching about technology through an historical and indigenous context may be the only realistic option.

Despite the severe difficulties surrounding its implementation, teachers and students in South Africa remain excited about the opportunities presented by doing technology.

Japan

Since 2002 Japanese lower secondary schools have offered industrial arts and homemaking education as a compulsory subject for all pupils irrespective of gender although, in practice, they are usually studied along stereotypical lines. At upper secondary level, about 25 per cent of students move to a form of vocational school, which has close and extensive contacts with industry, indeed some private vocational schools are specifically for just one company.

The technology curriculum is split into two areas: technology and manufacturing and information technology with topics as shown in Table 3.1.

An example of a new elective under 'Programming and instrumentation/control' would be 'mechatronics' or 'applied mechatronics' (a combination of mechanics and electronics) offered in a 'systems and control' curriculum, similar to technology and design in Northern Ireland.

Table 3.1 Technology topics in Japan

		Technology and manufacturing	Information technology
Compulsory	1	Roles of technologies	Different means of information exchange
	2	Design of manufactured articles	Basics of computers
	3	Material processing technology	Use of computers
	4	Working and repair of devices	Telecommunication networks
Elective	5	Energy conversion	Use of multimedia
	6	Cultivation of crops	Programming and instrumentation/control

Germany

The previously separate educational systems of the German Federal Republic and the German Democratic Republic are still working toward organising a unified system. The education system is funded by both federal and state governments, but is controlled by the 16 state governments. There are many similarities however because of federal regulations, state co-operation and national projects.

There are generally three types of secondary school: *Hauptschule* (general secondary school, apprenticeship preparation), *Realschule* (general comprehensive school) and *Gymnasium* (high school, university preparation) and about one-third of all pupils attend each type of school. Technology education is not common in the *Gymnasium* and is not compulsory in the *Realschule*, for example, it may be an alternative to a second foreign language.

The aims of technology education are to provide functional knowledge about technical devices and processes, to teach technology specific methodologies, for example creativity, co-operation and communication and to develop evaluation and assessment capabilities.

Not all state systems, however, have technology as a compulsory subject at all levels. Technology subjects may be called *Technik* (process and systems, consequences of technology), *technisches Werken* (skill development through making) or *Arbeitslehre* (careers, technology and economics).

In Schleswig-Holstein, the technology curriculum spans the lower secondary years and covers:

- machine and production technology
- transportation and traffic
- electrical engineering
- construction and the built environment
- supply and waste management
- information and communication.

The teaching of technology includes:

- instruction, how to do something
- design exercises
- manufacturing exercises, planning of the production process
- technological experiments
- technological analyses
- technological exploration, outside school
- technological assessment and evaluation.

The Programme for International Student Assessment (PISA) study has had a significant effect on German education. Conducted by researchers from the Organisation for Economic Co-operation and Development, the study compares the achievements of 32 countries in three subjects: reading, mathematics and science. When the results were first published in 2001, Germany was ranked in the bottom third. This led to the development of national standards in the core subjects and focused educational resources in these areas, to the exclusion of technology.

Australia

Arguably the most significant date in the history of technology education in Australia was 1987, when all state and federal ministers of education agreed on the national goals for schooling in Australia. As part of this, they declared that the curriculum was comprised of eight learning areas, one of which was technology.

This declaration had profound implications. First, in secondary schools, the subject areas from which technology education developed were located within the elective areas of the curriculum. The implication was that these subjects provided learning experiences relevant only for specific groups of students with particular interests or career destinations in mind. Indeed, some of these subjects were regarded by students and the community as relevant only to a particular gender. Second, in primary education, technology had not generally been part of school programmes, and primary teachers had little experience to draw on to develop programmes. The challenge for technology was to determine the learning experiences that are essential for all students and are unique to technology education or best undertaken within the area.

An attempt at a national curriculum was made in 1994, when a national project in technology education was completed in which all the states co-operated in the development of a statement of technology education and profiles of student activities illustrative of that statement. This level of co-operation between states had never been experienced before. Although the two documents generated from this project were not legislated, and so do not constitute a national syllabus, they did provide guidance for the direction of technology education throughout Australia.

The statement in technology provided a framework for curriculum development, and is divided into four strands of learning: a process strand (design, make and appraise) and three content strands – materials, information, systems. This strand framework was matrixed with eight levels (spanning K-12) to give a sequence of statements about what students should experience in technology from the beginning of elementary school to the end of secondary school.

A number of trends can be identified in technology education in Australia:

- the lagging behind of primary school developments in technology compared to those at secondary level
- recognition for a general type of technology education to be a core and compulsory subject for all students in lower secondary studies

- secondary schools increasingly offering vocational courses, while colleges of technical and further education (TAFE), which were the traditional location for vocational education, becoming more involved in general education.

The five states and two territories of Australia are educationally independent and until 2010 had different educational systems, although the basic structure of six or seven years of primary and five or six years of secondary schooling was common. However, in 2009, a process began to develop a national curriculum. Phase 1 (published in 2010) included the subjects English, mathematics, history and science; Phase 2 (development began in 2010) included geography, languages and the arts; and Phase 3 is listed as 'the remaining areas' and includes design and technology, health and physical education, ICT, economics, business, and civics and citizenship. It is obvious that the phases reflect the political priorities. For example, history is in Phase 1 because the prime minister of the time considered history to be important. As each phase has been developed, notional hours for the teaching of each subject have been allocated and it seems that there will be very little time left in the school week for the Phase 3 subjects.

There has been robust debate around the development of the subjects undertaken so far and the expectation is that strong debate will continue in technology. While it would seem that the development of a new national curriculum for a country in the early part of the twenty-first century would be an opportunity to reconceptualise the nature of education, maybe departing from the traditional subject silo structure and focusing on the knowledge and skills students need for the future, this has not been the case and, as in England, the subject statements have been quite traditional.

As the development of a national curriculum in technology begins it seems that, despite strong lobbying, design and technology and ICT will be developed as a single subject, 'technologies'. This is not seen as a positive development by technology educators and may confirm some of the confusion about the nature of design and technology. This link with ICT, its confirmed low status and the limited time that seems to be available for the teaching of technology represent threats to the future of technology education in Australia.

Israel

The main objectives of Israel's technology education are to provide both general knowledge and career specific training, to enable career mobility, to help understand and deal with the technological world, to provide guidelines for problem solving and creative thinking and to take responsibility for the impacts of technology.

Educational policy was reformulated in 1985, following committee reports into the state of technology education. The goals of the reforms were to:

- at primary level, expose all children to technological culture and computers

- at junior high level, to expose all students to technology, the arts and computers and practical education in various fields
- at high school level, to expand the tracking system in technology education, to expand the scientific base of technology education, to update technological subjects, defer specialisation from grade 10 to 11 and expose all students to technology education.

New approaches, reflected in new curricula, included a balance between university academic requirements and the needs of industry and the Defence Force and a shift from industrial processes in workshops to the simulation of technological processes.

In elementary and junior high schools the traditional subject of arts and crafts was replaced (in 1991) by a science and technology programme that is compulsory for all students, and taught by subject specialists. Science in a technological society is the primary subject, and includes the following topics: materials, structures properties and processes; energy and interaction; technological systems and products; information and communication; earth and the universe and organisms and ecosystems.

Science and technology is a compulsory junior high school subject. It consists of a compulsory core, and elective modules suggested by the school. The subject is knowledge-based, aimed to promote creativity and initiative, and designed to inform vocational choices.

Technology and science in high school consists of core and vocational subjects, with both groups being both academic and practical. The tendency seems to be that the core subjects have the higher academic content and an emphasised scientific basis and the vocational elective subjects are more practical. The electives include mechanics, electronics, woodwork, biotechnology, industry and management, design, fashion and maritime studies.

In order to introduce all students to some technology studies in the academic high schools, a subject 'Science and technology for all' is being developed. A core includes studies in energy, data and communication systems, materials and design, with electives in electronics, transport technology and medical technology.

Conclusion

As technology has been present in some countries and some schools for a long time, it is surprising that there is still no consensus as to what school technology should be, how pupils learn when they study it, and what are effective teaching strategies. As de Vries (2006: 5) says: 'Of course one cannot reasonably expect a new or drastically reformed school subject to result in concrete evidence of success in just 20 years. Yet, for several countries the fate of technology education depends on that.'

Technological literacy is a common goal of technology education. Objectives to achieve that broad goal include understanding the role of technology in society, the relationship between technology and the environment and the development of cognitive skills such as evaluating, inventing, innovating, problem solving, creativity and manipulative skills. Content is broad and variable, although common areas include materials, systems, structures, CAD/CAM, control and IT.

History indicates that there is often an increased focus on technology education during times of economic downturn, reflecting an expectation that there is a direct relationship between technology education and economic activity. For example, in Australia, the periods of economic downturn in the 1890s, 1930s and 1980s all resulted in evaluations of technology (or technical) education and significant revisions to the curriculum. It could also be argued that the current economic crisis provided one rationale for the movement toward a national curriculum in Australia and certainly is the stated rationale for the STEM curriculum developments in England and the USA.

Different stakeholders tend to have different rationales for the promotion of technology education:

- Technology educators tend to place an emphasis on the social and cultural importance of learning about technology.
- Governments and industry often have a primary interest in workforce planning and ensuring an adequate supply of people into certain careers and occupations.
- Parents tend to focus on the general educative merits of technology education, such as problem solving and creative thinking.
- Students want to enjoy what they do and see technology as a welcome change from a largely theoretical and abstract curriculum.

Consequently in promoting our subject, we need to tailor our advocacy to the audience we are addressing.

Developing a technology curriculum and specifying in a document that it should be taught is one thing, but what actually happens in schools may be very different. A US colleague recently lamented that there are many examples of good practice in US schools, but not enough and they are not generally visible to those who make curriculum decisions. Technology education might disappear in the USA as a result of a number of factors coming together:

- a debilitating shortage of technology teachers
- lack of a cohesive approach to technology education
- the learning area changing its name to engineering
- the STEM amalgamation being promoted
- science looking to re-invent itself by teaching applications as well as content
- science teachers having the opportunity to teach this area.

And these factors may also resonate with the experiences of other countries.

Yet in many countries, technology is challenging a number of traditional characteristics of schooling – the decontextualisation of knowledge, the primacy of the theoretical over the practical and the organisation of the curriculum along disciplinary lines. Some of the innovative trends that are obvious in a number of countries include a movement from:

teacher as information giver	to	teacher as facilitator of learning
teacher-controlled learning	to	teacher–learner partnership
teacher-centred learning	to	student-centred learning
time, age and group constraints	to	individualised learning
materials-based organisation	to	needs-based activity
product centred	to	process centred
elective area of study	to	core subject
social irrelevance	to	socially contextualised

Given the identification of these trends, there is also a great degree of diversity throughout the world in technology education. This diversity ranges from the absence of core technology education (Japan) to its compulsory study by all students (Israel), an instrumentalist approach (Finland) to a basically humanistic approach (Sweden), a focus on content (USA) to a focus on the process (England), an economic rationalist philosophy (Botswana, China) to a more liberal philosophy (Canada), a staged and well-supported implementation of change (New Zealand) to a rushed and largely unsuccessful initial implementation (England), integrated with other subjects (science in Israel, IT in Australia) or as a discrete subject (Scotland).

The nature of technology education developed within a country must be designed to serve that country's needs and builds on the unique history of technical education resulting in a relevant technology education programme.

Questions

1 Looking across the world, what would you wish to take and include in the technology of your country? Why?
2 What do you see as *missing* from the different curricula?
3 Will the different technology curricula deemed suitable for the twentieth century meet the needs of society in the twenty-first century?

References

Black, P. (1996) *Curricular approaches and models in technology education.* Paper presented at JISTEC'96, Technology education for a changing future: theory, policy and practice, Jerusalem, 8–11 January 1996.
CETIS (2011) Centre for School Technology Education, Linköpings University, www.liu.se/cetis/english/index_eng.shtml (accessed October 2011).

de Vries, M. (1994) 'Teacher education for technology education' in Galton, M. and Moon, B. (eds) *Handbook of teacher training in Europe: issues and trends*, London: The Council of Europe and David Fulton Ltd.

de Vries, M. (2006) 'Two decades of technology education in retrospect', in de Vries, M.J. and Mottier, I. (eds) *International handbook of technology education: reviewing the past twenty years*, Rotterdam: Sense.

Elshof, L. (2005) 'Teacher's interpretation of sustainable development', *International Journal of Technology & Design Education* 15(2): 173–178.

Elshof, L. (2006) 'Technological education and environmental sustainability: a critical examination of twenty years of Canadian practices and policies', in de Vries, M.J. and Mottier, I. (eds) *International handbook of technology education: reviewing the past twenty years*, Rotterdam: Sense.

Ginner, T. (2007) *Implementing technology education: not just a question of excellent steering documents*. Key note lecture, New Zealand Technology Education Conference, Auckland, New Zealand.

Hill, A.M. (2006) 'Reflections on twenty years of wandering through the pathways and forest of technological education in Ontario, Canada', in de Vries, M.J. and Mottier, I. (eds) *International handbook of technology education: reviewing the past twenty tears*, Rotterdam: Sense.

Keirl, S. (2007) 'The politics of technology curriculum', in Barlex, D. (ed.) *Design and technology for the next generation*, Whitchurch: Cliffe & Company.

Layton, D. (1993) *Technology's challenge to science education*, Milton Keynes: Open University Press.

McCormick, R. (1993) 'Technology education in the UK', in McCormick, R., Murphy, P. and Harrison, M. (eds) *Teaching and learning technology*, London: Addison-Wesley.

Pavlova, M. (2006) 'Comparing perspectives: comparative research in technology education', in de Vries, M.J. and Mottier, I. (eds) *International handbook of technology education: reviewing the past twenty years*, Rotterdam: Sense.

Petrina, S. (1998) 'The politics of research in technology education: a critical content and discourse analysis of the *Journal of Technology Education, Volumes 1–8*', *Journal of Technology Education* 10(1): 27–57.

Petrina, S. (2007) '2020 vision – on the politics of technology', in Barlex, D. (ed.) *Design and technology for the next generation*, Whitchurch: Cliffe & Company.

QCA (2007) *Design and technology programme of study for Key Stage 3 and attainment target*, London: HMSO.

Todd, R. (1991) 'The changing face of technology education in the United States', in Smith, J.S. (ed.) DATER 1991 Conference Proceedings, Loughborough: Department of Design and Technology, Loughborough University of Technology.

WAG (2008) *Design and technology in the National Curriculum for Wales*, Cardiff: Welsh Assembly Government.

Williams, A. (1993) 'Rationale for technology education in NSW secondary schools' unpublished masters' thesis, University of New South Wales.

Part II

Debates about design and technology

Chapter 4

Why is transition from primary to secondary school so difficult?

Cathy Growney

Introduction

The move from primary school to 'big school' usually creates feelings of excitement and anxiety in most pupils: excitement at feeling 'grown up', looking forward to new subjects, the vast array of resources in secondary school and the challenges that lie ahead; anxiety at the thought of new subjects, the vast array of resources and the challenges that lie ahead! Most pupils do not worry about the continuity of educational experience or progression in their learning; for them it is all new and exciting. Yet for many pupils, perhaps particularly in design and technology (D&T), it is the lack of continuity and progression that causes them to soon feel confused and possibly less interested in the subject. Why is this happening and what should we be doing about it?

Pupils make transfers from one year to the next throughout their education and even sometimes more than once a year. Whatever the stage of transfer, school learning should be a seamless activity. In England, continuity of education within primary school from early years and the foundation stage (EYFS, 3–5-year-olds), through Key Stage 1 (5–7-year-olds) and Key Stage 2 (7–11-year-olds) is usually fairly smooth as, for most pupils, it occurs within the same institution. This is also true for year groups within secondary schools (Key Stage 3 (11–14-year-olds) and Key Stage 4 (14–16-year-olds)). However, despite numerous initiatives to reduce the impact, the difficulties many pupils experience in making the transition from primary school to secondary remain a thorny issue. This chapter discusses D&T in both primary and secondary schools and how the transition is managed, with specific reference to D&T. In the discussion, the systems and strategies introduced by schools to improve continuity of learning and whether such investments can guarantee success are also considered.

Researchers in the field of D&T have contributed to debates about the primary–secondary divide and the regression in pupil attainment and development (for example see Eggleston 1967, 1993, 2001; Chadwick 1989, 1991; Pearson 1990; Heath 1992; Stables 1995; Benson 2000, 2009; Barnard et al. 2000; Waldon 2001). However, it has to be acknowledged that the research base, in relation to D&T, is relatively meagre.

Where we are now and how we got here

Chapter 1 describes how D&T was introduced into the primary curriculum in 1990 and the difficulties and issues initially encountered. While many of these early difficulties have been addressed, primary D&T in some schools still suffers from a limited understanding among staff and high numbers of teachers not sufficiently qualified or confident in D&T (Benson 2012). However, the majority of D&T primary practice is good (OfSTED 2011) and is characterised by being integrated with other topic work, for example a focus on 'pets' might lead to building a home for a pet, which benefits pupils because the learning is contextualised and relevant. This structure also makes it easier for pupils to transfer learning from other areas, such as science, mathematics and literacy, to their D&T work. Although activities are structured, pupils generally have time to 'play' and experiment with materials and designs, and they often work in groups and spend time discussing what they are doing, what might or might not work and why particular decisions are made. As Kimbell, Stables and Green (1996: 108) noted, primary teachers 'allow their pupils to experience technology'.

Pupils' experience of secondary school D&T is almost the complete opposite of this. The work is distinctively separate, and even within D&T work in the different material areas is separate, and it can be difficult to see the relevance or any links with other subject areas. The work is highly structured and teacher led, pupils are presented with design briefs, the resources and time available and they are guided in the processes and skills they use. They usually work independently, with fewer opportunities for discussion. It is hardly surprising, then, that pupils experience a discontinuity between their primary and secondary D&T, which often leads to a regression in their learning.

The transition of pupils between the primary and secondary education phases is a long-standing problem in the education system (Galton, Gray and Ruddock 1999, 2003). Initiatives to overcome the discontinuity experienced by pupils were initially limited to addressing administrative issues, such as the passing on of school records, and the social and pastoral needs of pupils. The National Curriculum in England, which brought D&T into the primary school curriculum, was first implemented in 1990 (DES/WO 1990) and one of the intentions of a nationally prescribed curriculum was to ensure more coherence in the educational experience of all pupils, yet over 10 years after its introduction Galton, Gray and Ruddock (2003: 106) noted that 'almost all schools concentrated on administrative matters [between primary and secondary school] or easing the social passage of pupils from primary to secondary school.' There were, then, still difficulties in the academic and curricular aspects of transition.

There was a particular difficulty with D&T. Design and technology, as it appeared on the National Curriculum in 1990, had evolved from secondary school craft subjects (see Eggleston 1993), so primary schools were unfamiliar with it (although so were some secondary schools) and did not fully understand it. In addition, while secondary colleagues were required to understand and implement this new subject alone, primary teachers had to understand and implement *all* the NC subjects (there were nine). The need to respond to so many new changes meant that the attention required to tackle D&T often did not take precedence.

In order to teach D&T well, teachers need full proficiency in the subject, that is, confidence in the knowledge, understanding and skills, both pedagogic and practical, that define the identity and nature of the subject and competence to convey it through their teaching. In reality, primary teachers were expected to teach D&T with insufficient or no in-service training. Kimbell, Stables and Green (1996) noted the disparity in professional development opportunities for learning about D&T courses for primary and secondary teachers, with primary teachers having far fewer opportunities. There were myriad early problems in primary D&T and these included inadequate resources, insignificant budgets and few experienced subject leaders to empower and motivate their colleagues.

Benson has repeatedly illustrated the reasons for the difficulties experienced by primary teachers, 'Unlike literacy and numeracy, design and technology is not a priority for the government, nor has it been identified as a core subject' (2000: 3). Benson points out that many primary school teachers have not had any formal training to teach D&T, either because they qualified before D&T was part of the primary curriculum or, when it was no longer a statutory requirement, it was not part of their teacher training curriculum. Even where it is part of initial training, the time given to learning about D&T is often short so teachers still feel unprepared to teach the subject. Benson argues the importance of providing D&T training during initial teacher education courses for all primary teachers. Given the way in which D&T was introduced to the primary school curriculum, the wide spectrum of primary D&T implementation in the early years was not altogether surprising.

Less than 10 years after the introduction of the National Curriculum, in 1998, the government in England thought it necessary to introduce national strategies into the primary school sector, to raise pupils' attainments in literacy and numeracy. This was followed by the introduction of the Key Stage 3 strategy (DfES 2004) for secondary schools. The strategy initially focused on literacy and numeracy but a year later strategies were introduced for the 'foundation subjects', of which D&T was one. The strategy for D&T included initiatives for a more unified curriculum across the primary–secondary divide with proposals for 'bridging units', which were projects intended to improve curriculum continuity, progression and assessment. These projects commenced in primary school during Year 6 and continued in the secondary school in Year 7.

In 2003 the Primary Framework Strategy was reviewed, again focusing on literacy and mathematics with no mention of D&T. A review of the primary curriculum published in 2009 (Rose 2009) made curriculum recommendations that would have put D&T into a learning area called 'scientific and technological understanding', but these recommendations were not implemented, due to a change of government in 2010. Similarly, a major review of primary education published in 2010 (Alexander 2010) also suggested a curriculum that includes 'science and technology'. Since then government policy in England seems to have been focused on secondary school pupils' attainment and the structure and governance of schools.

Given that the current structure of educational provision in the UK is unlikely to change, the issue of the primary to secondary transition continues to be of concern. Several strategies have been suggested for improving pupils' experience and learning across the primary–secondary transition and these will now be considered.

Primary–secondary school partnership

Primary–secondary school partnerships can be beneficial in:

- providing a forum in which teachers can collaborate
- helping teachers share expertise
- ensuring progression throughout the depth and breadth of the curriculum
- raising standards of capability and attainment, in both primary and secondary pupils.

Mike Ive, a former HMI inspector and subject advisor for D&T, reasoned the case for establishing strong working links between secondary and primary school teachers for the benefits of pupils' progress. He argued that they would help 'non-specialist' primary school teachers develop their expertise and help secondary teachers build on pupils' 'previous experience' (Ive 1995). Such links would also provide a way to address the cornerstone of Kimbell and Stables' (2008) findings, in that the strengths of the teaching and learning primary D&T experiences could be built on and elaborated by D&T specialists within the secondary sector.

Unfortunately, tensions between secondary teachers collaborating with primary colleagues have been well documented (Chadwick 1989, 1991; Benson 2009). These tensions are partly due to that fact that many partnerships are not mutually negotiated but are led by the secondary school on the premise that the secondary teachers' 'expertise' would be shared with the primary teachers in order to address the perceived weaknesses in primary teachers' subject proficiency. The partnerships, then, often focused on developing primary teachers' practical and technical skills. Typically, secondary teachers approach the relationship with a perceived 'expert' status, which leads to feelings of inferiority and lack of empowerment among the primary professionals. These feelings are exacerbated by primary teachers' self-doubts, which are created by lack of, or limited, professional training in D&T. As early as 1967 Eggleston reported a superior attitude from secondary teachers, who held the view that primary education was merely a preparation for secondary education. This view is less common now, but remnants still exist and these need to be dispelled so that secondary teachers can gain an insight into pupils' capabilities and previous D&T experience (OfSTED 2011) and the D&T

foundations pupils gain in their primary school can be extended through to secondary. The need to establish trust and mutual respect between primary–secondary colleagues has been acknowledged as of paramount importance in developing effective partnerships (Beckett and Growney 2009; Growney 2011).

One way in which primary–secondary partnerships have become more effective is in their sharing of pupil data. However, while data are shared about pupils' attainments in English and mathematics, and sometimes science, there is little evidence that information on attainment in D&T is shared. This is, as with many educational issues, not straightforward. Primary teachers argue that when they do provide data these are ignored. One research study found this to be true, that there is a disinclination in secondary schools to use the information provided from primary school about pupils' attainment (Schagen and Kerr 1999). It has also been asserted (Cunningham 2012) that the highly politicised environment in which primary schools operate lead to data that are less reliable, supporting secondary teachers' caution.

An effective primary–secondary partnership would allow for discussion of learning experiences, attainment and moderation of standards and could be of benefit to teachers and pupils alike. Pearson (1990) and Benson (2009) contend that primary–secondary partnerships should be increasingly reciprocal; they stress the importance of the secondary specialists learning lessons from the nature and style of primary D&T as well as primary colleagues developing knowledge and skills. Similarly, Dakers and Dow (2004: 118) argue that 'the greater the degree of knowledge of primary teachers of secondary courses (and vice versa) the greater the degree of coherence, continuity and progression that can be achieved.' However, OfSTED (2011: 5) report that two-way liaison has largely been missing nationally:

> Pupils' work in D&T from their primary schools was rarely built upon by the secondary schools ... Teachers planned the curriculum without reference to what had gone before. This lack of continuity led ... to weak curriculum planning at Key Stage 3. Pupils said they found projects ... in D&T easy and the nature of the work was pitched too low or duplicated earlier learning ... This did not challenge pupils sufficiently, particularly the most able.

OfSTED (2011) also found that partnerships did little to enable secondary school teachers to gain insight into children's capabilities and previous experience. As a result, much D&T work in secondary school is repetitive or insufficiently challenging, which impedes pupils' transition and impairs their enthusiasm and motivation.

Another important reason for improving primary–secondary liaison is the acknowledged problem of boys' underachievement. OfSTED (2011) reported that the attainment gap in D&T to be one of the widest of the NC subjects and had widened since their report in 2007. Yet, they also reveal that of the 30 primary schools visited they 'found no gaps between the performance of boys and girls in D&T' (OfSTED 2011: 50). This suggests that one of the strengths in the teaching of D&T in primary schools is its ability to engage boys and girls equally and this is an area in which secondary teachers could learn from their primary colleagues.

However, it is not enough simply to establish liaison between primary and secondary schools; there has to be effective communication and agreed strategies for improving the transition for pupils. One way in which this could be done is through shared curriculum planning; possibly starting with 'bridging units'.

Planning for cohesions and continuity

Design and technology curriculum planning often takes place in primary and secondary schools with little attention paid to 'the other'. But what should the relationship be? Is it incumbent on primary teachers to prepare pupils for their secondary experiences or is it the responsibility of secondary teachers to plan their curriculum with consideration to pupils' previous experiences in primary D&T? Ive (1999), Dakers and Dow (2004) and others argue that the secondary curriculum should build on the D&T foundations from primary school, but, in practice, most secondary school D&T curricula are planned around the NC programmes of study and the expertise within the school. Does this matter?

In the mid-1990s, when looking at the development of D&T on the National Curriculum, Stables (1995) and Kimbell, Stables and Green (1996) identified several polarised disparities between the D&T experiences of pupils in Year 6 and those in Year 7. Table 4.1 shows some of the differences they found.

Kimbell, Stables and Green (1996) found that projects in both Year 6 and Year 7 were taught well and had been carefully conceived, pupils were equally keen on the curriculum area and proud of their work. However, the teaching and learning approaches were so different that D&T might not be recognisable as the same subject. In Year 6, learning styles were primarily grounded in uncertainties that pupils worked out, typically they gained confidence in doing so and became increasingly autonomous. The teacher guided the pupils by capitalising on their tacit knowledge. In contrast, in Year 7 learning was more predictable and characteristically directed by the teachers, who instructed pupils in what to do and often when and how to do it. This resulted in the development of pupils' explicit knowledge but also a reduction in their self-confidence and an overdependence on the teacher. Galton, Gray and Ruddock (1999) found that pupils felt that their Year 7 work 'underestimates what they are capable of doing and achieving' and Kimbell, Stables and Green, (1996) argue that these glaring discontinuities are factors in impairing progress across the primary/secondary phases. Interestingly, Stables (1995: 161) illustrated how pupils' abilities to learn through open-ended discovery learning was put 'on ice' in Year 7 rather than developed, yet this autonomy is precisely what secondary teachers aim to develop in pupils as they prepare them for public examination work.

Table 4.1 Year 6 and Year 7 differences

	Year 6	Year 7
Teachers	Rarely have D&T training	Specifically trained in D&T and have opportunities to update specialist training
Teaching assistants and other adults	Have little if any specialist D&T training	Have access to D&T training; some TAs have D&T specialisms
Pupils	Typically work in pairs or small groups to encourage scaffolding of learning through talk	Typically work on their own individual projects and talk is more commonly whole class teacher led
Workspace	Is their familiar 'ordinary' classrooms with basic specialist equipment	Is specialist rooms, often with sophisticated facilities
Schemes of work	Pupils carry out D&T activities that are usually integrated with other work and in which D&T areas are integrated	Pupils often experience a carousel of projects throughout the year, each focusing on one material area in D&T
Projects	Usually open-ended; often negotiated through collaborative activity with class members and the teacher	Usually prescriptive; predetermined by the teacher and lessons are systematically task specific
Links with other areas	Projects are more likely to have a specific cross-curricular nature	Only tenuous cross-curricular links are made; there is little collaboration with subject specialists from other curriculum areas
Timetabling	Usually more flexibility in timetabling – projects can be concentrated in intensive periods e.g. consecutive days or afternoons	Time-limited sessions at dedicated times each week, sometimes short sessions (45–50 mins)
Materials	Not prescribed	Mostly fixed and specified
Designing and making	The user is at the centre of the project and referred to throughout; fairly even attention divided between designing and making activities; designing is carried out through materials, often through group discussions	The user is not referred to often; little emphasis on 'designing'; it is usually done individually, on paper in advance; strong focus on 'making' to develop pupils' skills with new facilities, tools and equipment
Role of the teacher	Progress chaser; pupils are facilitated and supported in their decision-making process for selecting appropriate activities	Pupils are directed or instructed to do pre-allocated activities

(adapted from Kimbell, Stables and Green 1996)

Although their original findings were made some time ago, revisiting their work 10 years later Kimbell and Stables (2008) found that little progress had been made. Expecting a 'progressive pathway towards procedural autonomy ... where older learners might be only very loosely controlled by the teacher ... We had expected to find some sort of progressive transition from teacher-control to learner autonomy, but instead we found two such continua: one in primary years (1–6) and a separate one in secondary years (7–11). Critically, we found a massive discontinuity' (2008: 224–225).

There is increasing awareness and research evidence that in the primary–secondary transition pupils lose some autonomy and self-confidence in the process of D&T. It could be argued that this then leads to lower levels of attainment in the secondary school. However, few teachers in either primary or secondary are responding to this. There are reasons, of course: lack of time being the main constraint. But if primary teachers' emphasis is on the progress of their pupils and secondary teachers' on their levels of attainment, talking to each other to find ways to improve continuity across the transition phase would be of great mutual benefit.

One effective way of addressing this issue is for teachers from both sectors to work together to plan and teach 'bridging units'. Primary teachers introduce a unit of work to pupils in Year 6 and this continues into Year 7, to be completed by work in the secondary school. There are difficulties, of course. Secondary schools usually take pupils from a number of different primary schools and cannot always work with all those schools in planning and teaching the bridging units. Time is a constraint for teachers in both sectors and work of this kind does take time.

However, there are benefits to the use of bridging units. They help pupils to see the similarities, and differences, in their primary and secondary experiences of D&T. They foster continuity and progression in their learning and this helps to motivate pupils. They also counter the need for secondary teachers to undertake 'baseline assessments' when pupils first arrive and help to ensure that they extend pupils' learning by building on their prior experience. Waldon (2001: 160) stresses the importance of pupils' previous experience being valued, suggesting that it is not necessary to assess 'every last detail of children's capabilities' and that Year 7 should start with real assignments and engaging activities rather than the repetition of basic skills work.

Where pupils fail to see the connections between their primary and secondary work, or where teachers fail to build on pupils' primary experiences, there is often a regression in pupils' learning in secondary school, which should be of concern to all teachers.

Addressing Year 7 regression

Lack of attention to the academic and curriculum aspects of transition is partly responsible for the acknowledged regression in the performance of pupils in the early stages of secondary school and pupils becoming demotivated in their learning.

OfSTED (2002) identified not only a significant loss in pupils' attainment during Year 7, their first year in secondary school, but also a decline in the quality of teaching from the high point of Year 6 gradually downwards to Year 9. Two years later, OfSTED reported that the weakness of continuity persisted (OfSTED 2004). There are a number of suggestions why this regression occurs.

One possible reason is that in lower secondary school most pupils are taught through a 'carousel' system, in which they spend a few weeks in each of the D&T areas in Year 7, then again in Year 8 and Year 9 (OfSTED 2011). The time spent in each area varies, but is usually between six and 12 weeks, depending on how many areas of D&T there are within the department. The rationale given for using this system is that pupils should be taught by specialists in each of the areas, and that they need to experience each area in depth in order to be well informed before they decide in which area to specialise for their examination. In addition, it is often a convenient way of timetabling to ensure that all pupils get equal access to the different areas within D&T.

However, the weaknesses of the system have been recognised: 'Evidence shows that teaching which involves the frequent rotation of pupils among different materials areas can result in unbalanced achievement and a lack of progression' (DfES 2004: 60). OfSTED (2011: 34) found that it 'disadvantages ... learning, particularly for students ... who had difficulty in connecting different aspects of the subject. It also made the tracking of students' achievement problematic.' The carousel system favours teachers' 'subject' specialisms over their 'teaching' specialisms, it risks losing the coherent identity of D&T as one subject and it risks depressing pupil standards by providing a lack of continuity and progression. For pupils, particularly in Year 7, the carousel experience means constantly changing teachers and classrooms, time-limited experiences and goals within each of the areas and this often leads to a loss of self-confidence as pupils become confused by the changing nature of the subject. They also are unable to see the universal nature of the subject.

An alternative approach to the curriculum would be for secondary teachers to teach pupils in at least two of the areas and many are qualified to do so. This would mean that pupils would experience fewer rotations and teachers could provide a better continuity of learning.

It is also possible that the regression pupils experience is not due to the Year 6 to Year 7 transition, but to changes in pupils' experiences in Year 5 and Year 6 in the primary school. In Year 6, in contrast to Year 5, much of pupils' teaching and learning experience is influenced by the Key Stage 2 standard attainment tests (SATs); these are end-of-year tests in literacy and numeracy on which individual pupils and schools, are judged. Galton, Gray and Ruddock (2003: ii) found that 'much of Year 6, in the run up to the tests (SATs), consists largely of revision with an emphasis on whole class direct instruction.' Kimbell and Stables (2008) argue that the more open-ended, pupil-led nature of D&T is an antidote to the prescriptive learning of literacy and numeracy, but in some schools the drive for literacy and numeracy learning and revision reduces the time available for D&T

or constrains the experience. Galton, Gray and Ruddock (2003) found that the devotion in Year 6 to teaching for the SATs and the concomitant didactic style of teaching, hindered the open-ended learning characteristic of D&T. They also found that the didactic teaching style permeated other, non-SAT, subjects and that time given to non-SAT subjects was limited.

So what can be done?

Dakers and Dow (2004) identified four issues for smoothing D&T transition from primary to secondary school, which can be summarised as involving:

1. primary teachers' subject proficiency of D&T
2. secondary teachers' confidence in their primary colleagues
3. effective liaison between the two sectors
4. assessment.

These are, of course, interlinked and developments in any one of the areas would lead to improvement in other areas. This chapter has discussed these issues and suggested some ways in which they could be addressed, including:

- improving the initial training of primary teachers to ensure sufficient preparation to teach D&T
- professional development opportunities for primary teachers to renew and update their D&T knowledge and skills
- professional development opportunities for primary teachers to develop as D&T subject leaders in the primary school; Ive believed this to be the 'single factor' that leads to effective improvements (Ive 1999: 17)
- the development of primary–secondary school partnerships, with time and resources allocated to allow teachers to observe in one another's classrooms to learn from one another's practice, plan and teach shared units of work, assess work together
- secondary teachers teaching pupils across the D&T specialist areas rather than pupils moving through the carousel system.

Another possibility would be for pupils to build a portfolio of D&T work in the primary school and take it with them to the secondary school. This would allow them to show what they have done, but does require secondary teachers to be familiar with primary D&T work and to value what the pupils bring and be able to adapt their own teaching in order to build on pupils' existing levels of knowledge and skill.

Some schools have already introduced some of these ideas and they are proving effective. They all require time, energy and commitment and, with so many other pressing concerns, these resources may not be available for D&T but whatever schools can do to ease the transition for pupils will be of benefit.

Conclusion

This is an important debate. If we want pupils to enjoy D&T, be motivated to learn and to make progress then we have to make sure that we provide a good learning experience for them – whatever phase of education they are in. The cohesion and progression of the D&T curriculum through the primary–secondary divide is considerably stronger now than it was when the subject was newly introduced, but clearly there is still room for improvement.

Questions

1. How does this debate manifest itself in schools you know – does pupil learning in D&T suffer in Year 6? Do pupils regress in Year 7?
2. What can you do to find out more about what your primary or secondary colleagues do?
3. What do you think your department or school could do to make the D&T transition a better experience for pupils, and teachers?

References

Alexander, R. (ed.) (2010) *Children, their world, their education. Final report and recommendations of the Cambridge Primary Review*, London: Routledge.

Barnard, J., Farrell, A., Mantell, J. and Waldon, A. (2000) 'Bridging the gap: smoothing the transition between primary and secondary D&T', Birmingham: CRIPT.

Beckett, H. and Growney, C. (2009) 'Making the difference – working in partnership' in Benson, C., Bailey, P., Lawson, S., Lunt, J. and Till, W. (eds) 7th International Primary Design & Technology Conference – Making the Difference Conference Proceedings, Birmingham, June.

Benson, C. (2000) 'Ensuring successful curriculum development in primary design and technology' in Eggleston, J. (ed.) *Teaching and learning design and technology: a guide to recent research and its application*, London: Continuum.

Benson, C. (2009) 'Working together: primary and secondary teacher liaison', in Bekker, A., Mottier, I. and de Vries, M.J. (eds) *Strengthening the position of technology education in the curriculum*, proceedings PATT-22 Conference, Delft, the Netherlands, 24–28 August.

Benson, C. (2012) 'The development of quality design and technology in English primary schools: issues and solutions' in Ginner, T., Hallström, J. and Hultén, M. (eds) *Technology education in the 21st century*, The PATT26 Conference, Stockholm, Sweden/Linköping: Linköping University/CETIS/KTH.

Chadwick, E. (1989) 'Continuity between primary and secondary phases in science, technology and maths – an action research project in Hampshire' in Smith, J.S. (ed.) DATER 89 Proceedings of the Second National Conference in Design & Technology Educational Research and Curriculum Development, Loughborough, Loughborough University of Technology, 8–9 September.

Chadwick, E. (1991) 'Issues of progression in primary' in Smith, J.S. (ed.) IDATER 91: International Conference on Design and Technology Educational Research and Curriculum Development, Loughborough, Loughborough University of Technology.

Cunningham, P. (2012) *Politics and the primary teacher*, Abingdon: Routledge.

Dakers, J. and Dow, W. (2004) 'The problem with transition in technology education: a Scottish perspective', *Journal of Design and Technology Education* 9(2): 116–124.

DES/WO (Department of Education and Science and Welsh Office) (1990) *Technology in the National Curriculum*, London: HMSO.

DfES (2004) *KS3 The National Strategy Foundation subjects: design and technology framework and training materials DfES 0971-2004*, London: HMSO.

Eggleston, J. (1967) 'The social context of the school' in Klein, G. and M. Marland (eds) (2004) *A vision for today: John Eggleston's writings on education*, Stoke-on-Trent: Trentham Books.

Eggleston, J. (1993) *The challenge for teachers*, London: Cassell.

Eggleston, J. (2001) *Teaching design and technology*, 3rd edn, Buckingham: Open University Press.

Galton, M., Gray, J. and Ruddock, J. (1999) *The impact of school transitions and transfers on pupil progress and attainment: DFEE Report 131*, Norwich: HMSO.

Galton, M., Gray, J. and Ruddock, J. (2003) 'Transfer and transitions in the middle years of schooling (7–14): continuities and discontinuities', *Learning University of Cambridge DfES Report 443*, Nottingham: HMSO.

Growney, C. (2011) 'A perspective on the learning partnerships between a secondary school and its local primaries' in Stables, K., Benson, C. and de Vries, M.J. (eds) Perspectives on learning in design & technology education, Proceedings PATT25: CRIPT8 Conference, London, July.

Heath, J. (1992) 'Easing the transition from KS2 to KS3 through work in the food area of technology' in Smith, J.S. (ed.) IDATER 92: International Conference on Design and Technology Educational Research and Curriculum Development, Loughborough, Loughborough University of Technology.

Ive, M. (1995) *Design and technology: characteristics of good practice in secondary schools*, London: HMSO.

Ive, M. (1999) 'The state of primary design and technology education in England' in Benson, C. and Till, W. (eds) CRIPT Second International Primary Design and Technology Conference Proceedings: Quality in the Making, Birmingham, CRIPT.

Kimbell, R. and Stables, K. (2008) *Researching design learning: issues and findings from two decades of research and development*, New York: Springer.

Kimbell, R., Stables, K. and Green, R. (1996) *Understanding practice in design and technology*, Buckingham: Open University Press.

OfSTED (2002) *Changing schools: an evaluation of the effectiveness of transfer arrangements at age 11 (HMI 550)*, London: HMSO.

OfSTED (2004) *Standards and quality 2002/03 – HMCI Annual Report*. Office for Standards in Education, London: HMSO.

OfSTED (2011) *Meeting technological challenges? Design and technology in schools 2007–10*, London: HMSO.

Pearson, F. (1990) 'Liaison audit/questionnaire for National Curriculum in design and technology between the primary and secondary school' in Smith, J.S. (ed.)

DATER 90 Proceedings of the Third National Conference in Design & Technology Educational Research and Curriculum Development, Loughborough, Loughborough University of Technology, September.

Rose, J. (2009) *Independent review of the primary curriculum: final report*, London: HMSO.

Schagen, S. and Kerr, D. (1999) *Bridging the gap? The National Curriculum and progression from primary to secondary school*, Slough: NFER.

Stables, K. (1995) 'Discontinuity in transition: pupils' experience of technology in year 6 and year 7', *International Journal of Technology and Design Education* 5(2): 157–169.

Waldon, A. (2001) 'Bridging the gap', *Journal of Design and Technology Education* 6(2): 158–160.

Chapter 5

Is design and technology about making or knowing?

Mike Martin and Gwyneth Owen-Jackson

Introduction

The school curriculum in most countries has developed from centuries' old traditions of subject disciplines, areas of knowledge. As a result, school education is usually thought of as pupils acquiring 'knowledge', such as mathematical formulae, the names and sequence of kings and queens and the periodic table. The twentieth century saw developments in school curricula, with the introduction of subjects in which pupils learnt skills, such as performance skills in drama. The antecedents of design and technology (D&T) fell into this category, with pupils learning practical skills in woodwork, metalwork, cookery and needlework. As the curriculum developed further, in response to social and industrial needs, other skills were introduced into pupil learning, for example research skills, problem-solving skills and social skills.

The subjects that preceded D&T were focused on teaching pupils practical skills, with little attention paid to developing their knowledge or understanding. For a number of reasons – changes in the nature of the subject, raising its academic profile, improved understanding of skill development – D&T has increasingly focused on pupils' learning of knowledge as well as skills. So, the key debate here is about the relative value given to skills and to knowledge. Should the subject have a strong emphasis on the development of skills, as it traditionally has? Or should there be more focus on developing pupils' knowledge and, if so, what knowledge? These questions reflect different philosophical viewpoints on the purpose of D&T education and go to the heart of what is important in the subject.

This debate is an important one because, in many western countries, subjects focused on knowledge are perceived as 'academic' and those focused on skills as 'practical'. This divide in itself is not important, but the fact remains that higher value is placed on academic subjects, which are deemed suitable only for the more intelligent pupils, while practical subjects are considered less academic, and suitable for the less able pupils. This can lead to pupils being encouraged, or discouraged, to study D&T.

This chapter looks at the nature of skills and of knowledge and considers the place of each on the D&T curriculum.

Putting the debate into context

As was stated earlier, and elsewhere in this book, D&T grew out of practical craft-based subjects, so its original focus was on developing pupils' manual dexterity. From the introduction of compulsory secondary education, through to the 1960s and 1970s, practical lessons in 'craft' subjects became firmly established in secondary schools in the UK (Penfold 1988). Domestic science, metalwork, technical drawing and woodwork were commonly found on the curricula of secondary modern schools, as they placed a greater emphasis on practical work than did grammar schools. This served to reinforce the view, which still survives, that less academic pupils are more suited to practical subjects than their academic peers. Throughout this period, the practical subjects were focused on developing pupils' acquisition of hand-craft skills, with older pupils being allowed to use some machine tools. Pupils were taught to follow recipes, understand garment patterns or read technical drawings, and the products they made remained fairly static, to the point where pupils were making items of little relevance to their own lives. An example of this was the teaching of forge work through the making of a poker for a coal or wood fire, something that may have had real purpose in the early part of the twentieth century but, 50 years later, was hard to justify.

Beginning in the 1960s, innovative practice in a few schools and the development of various curriculum projects saw the introduction of elements of design work, which developed new skills and knowledge in pupils (see Kimbell 2004). This influenced the thinking behind the creation of 'design and technology' on the National Curriculum in the UK in 1989 (see Chapter 1). This envisioned a subject in which pupils used knowledge in order to design and make products, pupils were to be educated to develop 'capability to operate effectively and creatively in the made world' (National Curriculum Design and Technology Working Group 1988). Unfortunately, this vision never really came to fruition.

There also appears to be a dichotomy in the current National Curriculum requirements for D&T (QCA 2007). The programme of study requires pupils to acquire knowledge, such as knowing how to join materials. Yet the attainment target, against which pupils are tested, refers only to the 'processes' of D&T, such as developing and clarifying ideas, planning, making, testing and evaluating. In the attainment target, the only reference to knowledge is the expectation that knowledge will be applied when making decisions about the materials, tools and techniques to use. This is similar to the situation in other countries (see Chapter 3) and it will be interesting to see if this dichotomy will be addressed in any future review.

This tension in NC requirements creates difficulties for teachers who are required both to teach the programme of study and test pupils against the attainment target. It seems that the National Curriculum requirements created great opportunities for this new subject while, at the same time, inhibiting its development.

Debates about design and technology

Following the introduction of the National Curriculum, D&T teaching continued to develop as new materials, new tools and new processes were developed in business and industry, and schools attempted to emulate these. The increasing accessibility and affordability of a range of materials also impacted on D&T, for example smart materials, 2D design packages, 3D design packages, PICs.

The content of the design and technology curriculum has expanded in order to accommodate these new developments, as shown in Table 5.1. This, we suggest, is untenable. As the available curriculum time reduces how can we continue to increase the skills and knowledge that we expect pupils to learn? Yet, if we are to reduce the curriculum, what do we omit? Are all the skills still necessary? Is all the knowledge still relevant?

Skills

As the subject grew out of traditional craft-based work, we will first consider the place of skills within the D&T curriculum. Many skills are developed in D&T work, but here we focus on the development of practical skills through the making of products.

If the D&T curriculum is to be skills focused, then we need to ask why do we teach skills and what skills are important? Making things is an inherently human activity that develops not only pupils' motor skills and hand–eye co-ordination but also a range of other skills and knowledge and helps to develop their sense of worth. As Barlex (2003: 4) notes:

Table 5.1 Skills and knowledge in the teaching of design and technology

Skills	Up to 1960s	1970s	1980s	1990s	Present	Future?
Hand tools	■	■	■	■	■	■
Machine tools		■	■	■	■	■
Drawing skills		■	■	■	■	■
Designing skills			■	■	■	■
2D CAD				■	■	■
3D CAD					■	■
Rapid prototyping						■
Knowledge						
Properties of materials	■	■	■	■	■	■
Materials processing	■	■		■	■	■
Manufacturing systems			■	■	■	■
Strategic knowledge			■	■	■	■
Technology and society				■	■	■

The real products of design and technology education are empowered youngsters, capable of taking projects from inception to delivery; creatively intervening to improve the made world; entrepreneurially managing their resources; capably integrating knowledge across domains, sensitively optimising the values of those concerned and confidently working alone and in teams.

This is an important reason for teaching skills; through learning practical skills pupils develop a wide range of other valuable skills and qualities, such as confidence, decision making, planning and managing. These skills and qualities contribute to the general development of pupils' education and to their development as citizens and workers. So, we would argue, it is important that D&T retains a skills focus in order to help pupils develop a wide range of skills and qualities. However, we also believe that this wider learning should be made more explicit to pupils so that they begin to see D&T as not only about 'what they make' but also about 'what they learn'.

This leads to the question, then, of which skills are important? Table 5.1 shows that, while new skills have been added to the curriculum, in order to meet changing social/economic needs, the 'old' skills have been retained and curriculum content has just grown. In order to decide which skills should be continued we need to look at what pupils learn when they are learning, for example, how to saw a piece of wood, solder an electronic component, stitch a seam or bake bread. The skills we teach should be the vehicle for the learning we want to take place, so the choice of skill should be dependent on the learning outcome. We accept that there needs to be much more research in this area in order for us to make informed decisions and we think that this is an important area for research. Our position, then, would be that we cannot dictate which skills should be taught but that teachers or schools should make their own decisions, informed by what they want the pupils to learn.

We also acknowledge that consideration needs to be given to the wider range of skills developed through learning in D&T, skills such as decision making, planning, applying values, working collaboratively. It is outside the scope of this chapter to consider these but we acknowledge that many valuable skills can be developed through a well-planned D&T curriculum.

We also think it is timely to give consideration to how we teach skills in D&T. The roots of the subject lie firmly in the craft-based traditions of blacksmithing, carpentry, cooking and needlecraft. In those crafts, pupils developed their skills through an apprenticeship model. Practical activities were observed, copied and undertaken with the 'master' over an extended period of time and the tasks progressively became more difficult. The pupil, or apprentice, eventually completed training and was assessed on the quality of their craft skills and the measure of success was the extent to which the 'novice' was able to achieve the same quality outcome as the 'master'. The pedagogy traditionally associated with teaching craft skills, the apprenticeship model, found its way into the teaching of

D&T craft skills. Teachers demonstrate to pupils the use of a tool or how to carry out a process, which they are then expected to remember (sometimes for a week, until the next lesson) and mimic as they use the tool or carry out the process themselves, in the same way that the teacher did it.

This transmission model of skills acquisition, from novice to master, has been superseded by other models of learning (see Speelman 2005) but many later theories draw on Dreyfus and Dreyfus (1986). We, therefore, focus our discussion on this model, which identifies different stages in skill development. The stages Dreyfus and Dreyfus proposed were hierarchical, from novice, through advanced beginner, competent, proficient to expert. This skill development, according to their model is characterised by increasing understanding of the context and the development of decision-making skills and so is worth considering in the discussion about the value of skills in D&T education.

Dreyfus and Dreyfus argue that, in developing skill expertise, the learner moves from following given rules and principles to acting intuitively in response to each situation. They suggest that these are step changes that require continued practice in real situations, and that 'training programmes' should be designed to support and encourage learners through these steps. In a later paper (Dreyfus 2004), they also suggested that the 'emotional response' of the learner is an important aspect of skill development. If a learner develops a skill and improves performance, or achieves a goal, this provides positive emotional reaction, which enhances further learning. Lack of emotional involvement, conversely, leads to boredom and regression.

This leads us to suggest that we reconsider how we teach skills to pupils. If we acknowledge where pupils are in the hierarchy of learning skills, for some they will be novices, for others they may be advanced beginners or competent, then we can better tailor teaching to develop their skills in such a way that we foster positive emotional responses and growth in their skill development.

Another theoretical approach which has been widely accepted is that of Kolb's learning cycle (Kolb 1984). Kolb proposed that, in order for effective learning to occur, there is a circular process, in which a real, practical, concrete experience leads to observation and reflection. This, in turn, leads to the development of understanding or abstract concepts, based on the concrete experience. This understanding, Kolb suggests, is then tested out in new situations, or new concrete experiences, and so the cycle continues.

In D&T, we often provide pupils with the 'concrete experience' but less often do we provide opportunities for them to reflect on their practical experience and, through reflection, develop their conceptual understanding. If we want to fully develop the educational value of skills learning we believe that more consideration should be given not only to which skills are taught but also how they are taught.

The development of D&T in the UK has seen a move away from a skills-focused curriculum to a knowledge-focused one, so we now consider the value of this.

Knowledge

Traditionally, the knowledge taught in D&T has been about the materials, tools and processes required for skill acquisition and development. For example, pupils have learnt the names and properties of different metals, woods, fabrics and food ingredients; they have learnt the names of the different ways in which these materials can be joined and combined and the names and uses of relevant tools and equipment. We would argue, though, that there is more to D&T knowledge than this.

First, however, we need to make clear that 'knowledge' can mean different things. Propositional, or declarative, knowledge (Ryle 1949) refers to factual knowledge, for example the nutritional content of an egg, the tensile strength of steel or how silk fabric is produced. Propositional knowledge is sometimes referred to as 'knowing that'. Propositional knowledge is linked to conceptual knowledge and understanding (McCormick 1997), in which items of knowledge are put together in a meaningful way and can then be transferred to new situations. This, we would argue, is the aim of D&T education, as McCormick states (1997: 143) conceptual knowledge 'is not simply factual knowledge, but consists of ideas that give some power to thinking about technological activity'.

Procedural knowledge, in contrast, is knowing how to do something, for example bake bread, thread a sewing machine or solder a component to a printed circuit board. McCormick (2002), citing Stevenson, suggests that procedural knowledge can be classified into levels, from the simplest level of knowing how to carry out a practical skill through to the higher level of what McCormick calls 'strategic procedural knowledge' (2002: 96). Strategic procedural knowledge, referred to by Hope (2000) as 'know relevance', requires not only knowledge of procedures but also knowing which procedure to choose and when to do it.

While propositional knowledge and procedural knowledge are linked there are clearly differences, for example knowing that metals can be joined by using heat is very different from knowing how to braze and weld. Similarly, knowing the recipe for cheese sauce is one thing but being able to produce a dish of cauliflower cheese for four people to eat at a given time is quite a different thing. McCormick (2002: 104) argued that the linking of conceptual knowledge with procedural knowledge, which he linked to 'qualitative knowledge' (the 'rule of thumb knowledge applied by experts), is central to design and technology.

There is another type of knowledge that is important in D&T and that is tacit knowledge (Polanyi 1967). This refers to knowledge that is hard to quantify or describe and is often not explicit or acknowledged. Tacit knowledge comes into play, for example, when you know how a fabric will drape or when you recognise that the pillar drill has the wrong sized drill because of the sound it makes. Experienced practitioners can use this tacit knowledge but find it hard to explain what it is they know or how they know it. Such knowledge is essential in gaining mastery of materials and processes.

As with skills, the knowledge taught in D&T has grown as new aspects have been added, but with 'old' aspects of knowledge remaining. As Kimbell and Perry (2001: 8) wrote more than 10 years ago 'technological knowledge is currently exploding exponentially' and we believe that this is more apposite today with electronic communication making more knowledge even more readily available. This means that we cannot keep adding to the list of propositional knowledge that we teach pupils.

Perhaps now it is time to consider what knowledge is sufficient and appropriate for pupils in the twenty-first century. Over 30 years ago the Department for Education and Science (DES) noted that pupils did not necessarily need a body of knowledge but they did need 'to know what to find out, what form the knowledge should take, and what depth of knowledge is required for a particular purpose' (DES 1981: 12). Perhaps it is time to revisit this idea and consider how we can design learning experiences in which pupils learn how to recognise what knowledge they need, how to find knowledge, how to evaluate knowledge and how to appropriately apply the knowledge they have. Is it still important for pupils to know the properties of specific materials or should we teach them how to look up what they need to know and how to critically evaluate information they find? In knowing about and using the materials, tools and processes, should pupils not give thought to the values inherent in their use? Do pupils understand the consequences of selecting to use certain materials for their products or the impact of recycling materials on economies elsewhere? While these aspects of knowledge are mentioned in National Curriculum documents and examination specifications, it is our view that less attention is paid to them than to other aspects.

We do not want to assume that we know what the fundamental principles of design and technology are and we cannot offer a list of what we think D&T subject knowledge should include. In fact, as McCormick notes (1997: 144), the multidisciplinary nature of most D&T work 'makes defining a knowledge base, and the search for a unique common set of procedures or concepts, particularly difficult'. However, we do think that (in England) the current conception of D&T requires a radical overhaul. The knowledge currently being taught, we suggest, is not sufficiently relevant, engaging or useful to pupils and, in our view, often not sufficiently intellectually challenging.

Skills and knowledge

As you would expect, good D&T teaching requires a judicious combination of both skills and knowledge. Owen-Jackson and Steeg (2007: 172) suggest that technical knowledge includes propositional knowledge and procedural knowledge and that both are necessary and relevant. Skills need to be informed by knowledge and through the processes of using skills new knowledge is developed.

McCormick (1997) argues convincingly that D&T teaching should focus on enabling pupils to develop their 'strategic knowledge' by allowing them to make decisions about when and how they draw on and apply their propositional and procedural knowledge. Most D&T teaching does the opposite of this. You are likely to be familiar with schemes of work structured so that pupils first identify the problem or need, generate possible solutions, develop one solution through to making then evaluate it. This form of teaching allows no space for pupils to make decisions about what knowledge is needed or when to undertake different activities. It would be more challenging, we accept, for pupils to have this degree of freedom and control but if we are concerned about developing pupils who are knowledgeable about, and can be effective in, the made world then we have to face up to this challenge.

There is also a growing literature about 'embodied cognition', the notion that the body and the practical activities we undertake shape our thinking (Wilson 2002; Anderson 2003; Lakoff and Johnson 2003). Developing knowledge through doing is not a new idea and was written about almost 80 years ago (Dewey 1933). Papert (1980, 1994) provided further support for learning through making, suggesting a theory of learning he called 'constructionism' in which children learnt while making things. More recently, Claxton, Lucas and Webster (2010: 9), writing about embodied cognition within education wrote that 'Handling materials and literally "getting to grips" with problems allows us to see things and inform our understanding in ways that simply looking and thinking do not.' They suggest that 'academic' learning needs to become more practical and not the other way around! They also argue that 'practical learning' is cognitively demanding, involves higher order thinking skills and that all learners would benefit from it.

Conclusion

We hope that we have shown in this chapter that, not surprisingly, focusing on either skills or knowledge separately is not helpful as it is the combined effect of both working together that supports successful designing and making.

It can be seen from what has been written that design and technology is a dynamic subject that will continue to change in response to the changing technological practice of professionals and industry and the changing social/economic context. Is it possible, therefore, to define the 'facts, concepts, principles and fundamental operations' (DFE 2011) inherent in the subject? We believe that, while materials and processes may change, there are important concepts and principles that can be defined. It is incumbent on D&T to teach these fundamental principles using the skills and knowledge relevant to the time.

Questions

1   What has been the balance of skills and knowledge in schools that you have experience of?

2   What skills, and what knowledge, do you think it is important for pupils to learn?
3   How do you think D&T teaching can change its focus from what pupils do to what they learn?

References

Anderson, M.L. (2003) 'Embodied cognition: a field guide', *Artificial Intelligence* 149: 91–130; www.agcognition.org/papers/AI_Review.pdf (accessed 30 March 2012).

Barlex, D. (2003) *Building on success: the unique contribution of design and technology. A report to ministers from the Design and Technology Strategy Group*, London: DfES.

Claxton, G., Lucas, B. and Webster, R. (2010) *Bodies of knowledge. How the learning sciences could transform practical and vocational education*, London: Edge Foundation.

DES (Department of Education and Science) (1981) *Understanding design and technology*, London: HMSO.

Dewey, J. (1929) *The quest for certainty: a study of the relationship of knowledge and action*, New York: Minton, Balch & Co.

DFE (2011) *Michael Gove to Twyford Church of England High School*. Speech available at http://www.education.gov.uk/inthenews/speeches/a0073212/michael-gove-to-twyford-church-of-england-high-school.

Dreyfus, H. and Dreyfus, S. (1986) *Mind over machine; the power of human intuition and expertise in the era of the computer*, Oxford: Blackwell.

Dreyfus, S.E. (2004) 'The five stage model of adult skill acquisition', *Bulletin of Science Technology & Society* 24: 177.

Hope, G. (2000) 'Beyond "draw one & make it"' in Kimbell, R. (ed.) Conference proceedings: Design and Technology International Millennium Conference, Wellesbourne, Design & Technology Association.

Kimbell, R. (2004) 'Design and technology' in White, J. (ed.) *Rethinking the school curriculum: values, aims and purposes*, London: RoutledgeFalmer.

Kimbell, R. and Perry, D. (2001) *Design and technology in a knowledge economy*, London: Engineering Council.

Kolb, D.A. (1984) *Experiential learning: experience as a source of learning and development*, Upper Saddle River, NJ: Prentice-Hall.

Lakoff, G. and Johnson, S. (2003) *Metaphors we live by*, Chicago: University of Chicago Press.

McCormick, R. (1997) 'Conceptual and procedural knowledge', *International Journal of Technology and Design Education* 7: 141–159.

McCormick, R. (2002) 'Capability lost and found?' in Owen-Jackson, G. (ed.) *Teaching design and technology in secondary schools: a reader*, London: RoutledgeFalmer.

National Curriculum Design & Technology Working Group (1988) *Technology in the National Curriculum: interim report*, London: DES/HMSO.

Owen-Jackson, G. and Steeg, T. (2007) 'The role of technical knowledge in design & technolgoy' in Barlex, D. (ed.) *Design & technology for the next generation*, Whitchurch: Cliffe & Company.

Papert, S. (1980) *Mindstorms: children, computers and powerful ideas*, New York: Basic Books.

Papert, S. (1994) *The children's machine: rethinking school in the age of the computer*, Hemel Hempstead: Harvester Wheatsheaf.

Penfold, D. (1988) *Craft, design and technology: past, present and future*, Stoke-on-Trent: Trentham Books.

Polanyi, M. (1967) *The tacit dimension*, London: Routledge & Kegan Paul.

QCA (2007) *Design and technology programme of study for Key Stage 3 and attainment target*, London: HMSO.

Ryle, G. (1949) *The concept of mind*, Harmondsworth: Penguin.

Speelman, C. (2005) 'Skill acquisition: history, questions, and theories' in Speelman, C. & Kirsner, K. *Beyond the learning curve*, Oxford: Oxford University Press.

Wilson, M. (2002) 'Six views of embodied cognition', *Psychonomic Bulletin & Review* 9(4): 625-636; http://people.ucsc.edu/~mlwilson/publications/Embodied_Cog_PBR.pdf (accessed 30 March 2012).

Chapter 6

Design and technology education: vocational or academic?
A case of yin and yang

Gary O'Sullivan

Introduction

The notions of yin and yang are heavily rooted in Chinese philosophy. Most people educated in the west think of them as polar opposites but after a lifetime of Chinese martial arts study, I have come to think of them as complementary. They are interconnected and interdependent, the image itself contains both light and dark, one feeds the other. The key is finding balance and harmony so both can be developed.

A similar (mis)representation can be seen in the development and implementation of design and technology (D&T) education. There are two differing philosophical positions surrounding the function and purpose of D&T education that could be seen, as with yin and yang, as polar opposites. These two positions regard D&T education as either vocational or academic. Academic education is generally taken to refer to the 'classical' subjects: mathematics, literature, languages, science, history. Vocational education, in contrast, is concerned with developing knowledge and skills appropriate for employment and has, over the years, included car mechanics, hairdressing, business skills. Design and technology education can be seen as both a vocational and an academic subject. As a vocational subject it is associated with notions of both craft and employment preparation. As an academic subject, a greater emphasis is placed on the nature of technology and technology in a critical social context. These two paradigms are reflected in the yin and yang symbol.

However, this perceived dualism has dogged the progression and position of D&T over a number of years and there has been little consideration of interconnectedness, harmony or balance. This polarisation could, in fact, see the demise of the subject as a core component of a broad general education. This chapter will briefly consider the historical context of the vocational–academic debate, the purpose of education and the role of design and technology in pupils' education.

Historical context

There has long been a vocational–academic divide in the UK education system. The modern education system in England and Wales was founded on the 1944 (Butler) Education Act, with similar Acts in Scotland (1945) and Northern Ireland (1947) (Gillard 2011). Although the 1944 Act is often attributed with introducing the tripartite system of schooling, with three types of secondary school: grammar, secondary modern and technical, it was, in fact, the Labour government of the time that was responsible for this (Benn and Chitty 1996). Grammar schools were open to academically 'able' pupils who passed an examination (the 11+) taken at 11 years of age, at the end of their primary schooling. The grammar school curriculum was academic, intended to prepare pupils for university and professional careers. Secondary modern schools were for the less academic pupils who failed the 11+ examination, they focused on providing a basic education and practical subjects. Technical schools were intended for academically able pupils who wanted a technical education; they combined both academic and practical learning. In practice, however, the technical schools failed to materialise in significant numbers, partly attributed to a lack of suitably qualified teachers, so their effectiveness was never properly evaluated. This meant that pupils were mostly divided between academic grammar schools and practical secondary modern schools. As grammar school places were limited to the top 20 per cent of pupils, at the insistence of the government (Gillard 2011), the majority of pupils attended secondary modern schools.

One major concern with the tripartite system was that the schools were socially divisive. Although the intention was that the 11+ examination allowed grammar school entry to those pupils who proved themselves to be academically able, many grammar schools were located in more affluent areas while secondary moderns were not. This provided support for the belief that grammar schools were for pupils from the middle to upper social classes and secondary modern schools were for pupils from the lower middle to lower classes. Some regions in England still retain the tripartite system although many moved to comprehensive schooling in the 1960s and 1970s.

The school system, however, continues to change and the UK government has introduced different types of school including, for 14–19-year-olds, studio schools and university technical colleges (UTCs). Studio schools focus their curriculum on 'project-based' learning and integrate academic and practical education to prepare pupils for work. Numbers of studio schools are very small but of those open, or preparing to open, some are focusing on areas related to D&T, for example International Food and Travel Studio, Da Vinci Studio of Science and Engineering, Midland Studio School focusing on Engineering and

Health and Social Care (http://media.education.gov.uk/assets/files/list%20 of%20studio%20schools.pdf (accessed 3 June 2012)).

University technical colleges are schools that specialise in teaching technical subjects, together with business and IT skills. Every UTC is sponsored by an industrial or commercial company and a university and its curriculum focuses on meeting local needs, likely to be the local employment and industry needs. The number of UTCs is small but growing, and they all specialise in disciplines related to some aspect of technology, engineering, manufacturing and/or design. The government has said that the UTCs would be 'prestigious' and would raise the status of technical education, but there are some concerns that they would lead to a 'two-tier' socially divisive school system such as existed with grammar and secondary modern schools (BBC news, 4 April 2010 http://news.bbc.co.uk/1/ hi/education/8602723.stm (accessed 6 June 2012)).

Other countries also have divided education systems, with different types of secondary schooling available for pupils with different aptitudes and interests, usually separated according to academic ability. Many European countries, for example, have academic, gymnasia, schools and practical or technical schools based on occupations. In Finland, which has a good reputation for its educational system, upper secondary schooling is optional but most pupils attend and can select academic or vocational school, almost half attend vocational schools (Business Insider International 14 December 2011 http://www.businessinsider. com/finland-education-school-2011-12?op=1 (accessed 6 June 2012)).

Such divided education systems also contribute to the notion of two types of learning: academic and vocational, the yin and the yang. The link between grammar schools and social status has meant that, in the UK, academic learning has higher value and status while vocational education, and by association D&T education, has always suffered from the perception of being not academic and so having low status (Evans 2008). This is not, however, true for all countries although there often are marked divisions.

Vocational education and training (VET) has a turbulent history, swinging in and out of favour depending on the policy whims of the governing political parties. Nowhere has this turbulence been more evident than in England (see Hayward et al. 2009). And it was, again, a Labour government in the 1970s that argued for the need for a stronger vocational element in schools. This was promoted by the Prime Minister at the time, James Callaghan, in a speech at Ruskin College in 1976 which led to the 'Great Debate' in education. This debate was about the school curriculum and the role of education in helping Britain to be economically and industrially successful. In some ways, the legacy of this debate can still be felt in education today.

Following the debate, a number of initiatives, each designed to support and develop vocational education in the school curriculum for 14–18-year-olds, have come and gone. These began in the mid-1980s with the Technical and Vocational Education Initiative (TVEI), a curriculum focused on preparation for work, including work experience. This scheme was the government's response to high youth unemployment and employers' claims that schools did not properly prepare young people for employment. The initiative was well supported by government, at a cost of £900 million funded by the Department of Employment (*not* Education), and was developed mainly by politicians with little input from educationalists. Under the initiative, schools and colleges worked together to teach a curriculum that had a vocational focus but also included general academic knowledge and skills. There was also a requirement for every pupil to have some kind of 'work experience', requiring schools to negotiate with a range of local businesses and employers. The Initiative ran for a number of years in school but was impacted by the introduction of the National Curriculum and the changing environment and was formally ended in 1997.

In 1993, as the TVEI was fading, general national vocational qualifications (GNVQs) were introduced and continued until 2007. GNVQs were vocational qualifications for school pupils, offered in a number of employment areas, and were intended to bring 'parity of esteem' to academic and vocational qualifications. The GNVQ most closely related to D&T was that of 'manufacturing'. This was adopted in many schools and could embrace any or all of the areas within D&T: food, resistant materials, textiles. Some food technology departments contributed to the teaching of health and social care as this covered food and nutrition.

Initially aimed at 16–18 year olds, a new GNVQ qualification, Part One, was developed for 14–16-year-olds in school. All post-16 GNVQs were available at three levels, with foundation regarded as the equivalent of low-level academic attainment at 16 years old, intermediate as high-level academic attainment at 16 years old and advanced as equivalent to A level attainment at 18 years old, in theory appropriate for entry to university although universities were generally suspicious of these 'new' qualifications. Pupils could progress through the levels, as they could with academic qualifications. In contrast to TVEI, each GNVQ had a detailed specification and outcomes-based assessment criteria for which pupils had to provide evidence. Secondary school teachers teaching GNVQ had to acquire accreditation as an 'assessor' in order to be able to assess their own pupils. It has been stated that schools responded well to GNVQs and that they were studied by high numbers of pupils, many of whom stayed at school beyond 16 (Evans, undated). Despite their relative success, GNVQs were phased out, to be replaced by vocational GCSE and A level examinations, in 2002.

Vocational examinations were a further attempt to bring parity of esteem to vocational and academic qualifications, this time by putting vocational qualifications under the same umbrella as academic ones. Vocational courses were available in a range of subjects, and at different levels, again with the purpose of preparing young people for employment. For D&T, manufacturing and health and social care were joined by engineering, although this was not taken up by many schools mainly due to a lack of appropriate facilities and suitably qualified staff. Vocational examinations, in turn, were replaced by diplomas.

The Tomlinson Report (2004) was commissioned by the government as a review of 14–19 curriculum provision in response to concerns about pupil disaffection, many pupils leaving education at the age of 16, low levels of literacy, numeracy and IT skills among some pupils and to improve the quality and status of vocational education. Tomlinson (2004: 79) criticised the academic–vocational divide:

> There is no absolute distinction between vocational and general (or academic) learning. Good vocational provision develops skills, knowledge and attributes that are desirable in adult life generally, and not only in the workplace; conversely, much of what is learnt in general or academic learning is relevant to employment.

The report then recommended a unified programme of learning and qualifications for all pupils, the government did not implement all the recommendations of the Tomlinson Report but did introduce vocational diplomas.

Diplomas are available in a number of employment areas; those most relevant to D&T include construction and the built environment, engineering, manufacturing and product design and hospitality. The curriculum for each diploma was developed by partnerships in which employers and universities were heavily represented. They combine academic and vocational learning, but have a complex structure and are not being taken by large numbers of pupils. Although officially still available, diplomas are not on the curriculum in many schools and the proposal to extend diplomas to include 'academic' subjects, such as science and modern foreign languages, has now been shelved.

The 2011 curriculum review in England reconsidered the status and intention of design and technology education. The coalition government of the time stated that English, mathematics, science and physical education must remain compulsory for children of all ages. The status of all other subjects, including design and technology, was put under review and the government has delayed and delayed on announcing any decision – which perhaps indicates some of the struggles around defining a national curriculum. However, it looks as if there may be three possible options for D&T: an academic route, based on assessments through the international baccalaureate or an A level (higher level examinations), a vocational route based on technical education or a vocational route based on practical education. Whatever decision is made it seems that both general education and design and technology education are under significant threat!

The purpose of education

One of the reasons for the continuing debate about the academic and vocational aspects of education is that it is not clear, certainly in the UK, what the purpose is of secondary school education.

Historically, there seems to have been more certainty. In 1864, the Schools Inquiry Commission reviewed education provision at secondary level and made recommendations that reflected the class structure of the period. Three grades of schooling were identified: first grade to prepare upper middle-class boys for university, second grade to prepare middle-class boys for the civil service and the army, third grade for lower middle-class boys to facilitate their entry into the workforce as tradesmen. Twenty years later, the Royal Commission on Technical Instruction produced two reports between 1881 and 1884 where it was noted that the middle classes were disadvantaged because of their lack of exposure to technical study (see Spens 1938). Around 1882 the Samuelson Commission recommended handicraft be introduced into schools in order to arrest Britain's economic decline.

According to the School Board of London (1898) handicraft was offered to develop practical hand-craft skills in those pupils who were deemed to be struggling with brain work. In 1913 the Board of Education created a new category of 'junior technical schools'. Gillard (2011) identifies these as providing two- or three-year post-elementary courses for both boys and girls, bringing together general and vocational education to prepare children for industrial employment at the age of 15 or 16 years.

Similarly, politicians implementing the 1944 Education Act, discussed earlier, regarded schooling as providing an academic education for 'able' pupils and vocational preparation for the less able. During the 1960s the tripartite system was phased out, to be replaced with comprehensive schools that all pupils attended, irrespective of academic ability. The curriculum of each comprehensive school depended to some extent on its history, that is, whether it had previously been a grammar or a secondary modern school.

Callaghan's Ruskin College speech, in 1976 raised questions over the effectiveness of the comprehensive curriculum in meeting the needs of the British economy. In 1977 the government published the Green Paper, 'Education in schools', which argued:

> It is vital to Britain's economic recovery and standard of living that the performance of the manufacturing industry is improved and that the whole range of government policies, including education, contribute as much as possible to improving industrial performance and thereby increasing national wealth.
>
> (Green Paper Education in Schools: A Consultative Document 1977: 6)

It is clear from this that, during a period of rising unemployment and major decline of Britain's economy, there was concern that education was failing to meet the needs of society. This perceived failure gave rise to a significant change in emphasis for education, which was called 'new vocationalism'. This term refers to a particular political ideological shift from education being a means of reducing inequality to it having a functional purpose, to meet the needs of the economy, particularly industry and business. This purpose seems to still be present in the secondary education curriculum today.

Economic instrumentalists, a strong lobby group that have traditionally been behind the development of D&T education, believe that education has an important role to play in the economic competitiveness of nations, where the creation of wealth should be the overriding aim. One method, they would suggest, for achieving this is through vocational education delivered, in part at least, via D&T education. Saunders (2000) argues that this occurs through selection and preparation of pupils, i.e. funnelling some pupils down a pre-employment pathway thus helping to underpin 'new vocationalism' and the introduction of education as training.

The liberal perspective of education stands in opposition to 'new vocationalism'. The liberal perspective regards education as important in its own right, rather than to fulfil some extrinsic factor such as employment or economic objectives, that 'general education' is suitable for all aspects of future life, including work and that vocational preparation is best undertaken either at work or just prior to beginning work. Liberal educationalists believe that general education that is broad, deep and informed by the whole culture is the best preparation for life. Such education may include interactions with the world of work, but as a pedagogical process rather than direct preparation for a particular occupation. The effectiveness of education, from this perspective, should not be narrowly analysed by relating it to one particular employment or even a specific economic agenda:

> What is important for this perspective is the democratic imperative that no child should be denied access to these forms of knowledge and experience in the mistaken belief that they are not 'relevant' either to them or an extrinsic need like that of employers.
>
> (Saunders, 2000: 692)

Saunders identifies a major problem with this approach, that of finding ways for all pupils to access such an education. The way in which education is structured in many countries means that those with access to wide and deep cultural 'funds of knowledge' generally fare better than those with limited access. According to Bereiter (2002), liberal education gives all learners access to a culture that transcends the particularities of their social and ethnic backgrounds. This liberal perspective on education supports D&T within a general, rather than a vocational, education.

In England, because of concerns about Britain's manufacturing and economic performance and a 'failure to provide young people with a proper practical and technical education' (Wolf 2011: 4), the government commissioned a review of vocational education provision. The Wolf Report on 14–19 vocational education (2011) was critical of much vocational provision and among its recommendations was that children should study mainly academic subjects until the age of 16 and, if they do not attain good examination results in English and mathematics by that age, they should continue to study those subjects. It further recommended that vocational qualifications should be of high quality, and ways in which this might be achieved included the involvement of employers in quality assurance and assessment. Following the publication of the Wolf Report (2011), the Education Secretary in England stated that the qualifications system is unfair for pupils and is harming the economy. He claims that vocational education, and the increase in 'pseudo-academic' courses, is very damaging and he has reduced drastically the number of vocational courses available to school pupils.

The Wolf Report brought VET to the forefront of educational debate. These debates are framed by wider social issues; for example in England, there are high levels of unemployment and disaffection, particularly among young people. The economy has been in disarray following the banking crisis of 2008, and in 2011 riots occurred across Britain involving many young people. Proposals to raise the education participation age to 17 in 2013 and to 18 in 2015, have done little to appease the unrest. Internationally, policymakers charged with developing education and employment strategies usually target the 14–19 age sectors for their reforms. Often these reforms are too narrowly focused on VET initiatives designed to fulfil an economic instrumentalist agenda:

> Little makes more difference to people's lives than the empowerment they receive from education. But for those young people whose aptitudes and talents are practical, expectations are too often limited and opportunities restricted. For far too long vocational learning has been seen as the poor relation of academic learning.
>
> (Hayes in Wolf 2011: 17)

Some countries have shown that it is possible to focus on developing pupils' general technological expertise, rather than specific vocational knowledge and skills, and still make economic and educational progress. For example, over the past decade Finland has consistently ranked highly in both educational and economic assessments. The educational success is achieved consistently across all schools irrespective of pupils' background or socioeconomic status. Autio and Hansen (2002) suggest that Finland's highly practical technology education is responsible for its innovations in mobile telecommunications, one source of its economic success. They argue that outcomes from learning technology include individual responsibility, creativity, perseverance, initiative and a positive picture of oneself, and they suggest that self-esteem is built on practical rather than academic achievement. These are strong arguments for D&T as a general academic subject.

Academic or vocational? The yin and the yang

So what is the role and purpose of D&T in all of this? Is D&T a subject that can contribute to pupils' general education, providing them with knowledge and skills whatever path they take or is its purpose to provide the necessary labour requirements to meet an economic instrumentalist agenda?

Historically, the subjects that now form D&T education have contributed to pupils' vocational education: craft skills were taught to enable boys to work as tradesmen or labourers, and domestic skills were taught to girls to train them for paid domestic work or housewifery. As social and economic needs changed the purpose of D&T education changed, first to educating pupils in 'life skills' then to contributing to their general academic learning.

The aim of D&T education in some curricula is to support the development of students' critical technological literacy, which will serve them well as informed and empowered citizens. Critical technological literacy is seen by some as more important and much broader than functional literacy (Davies Burns 2000; Petrina 2000; Compton and Harwood 2006) and many contemporary D&T curricula around the globe (see Chapter 3) focus on developing pupils' technological knowledge, technological literacy or the development, through technology, of general knowledge and skills. In New Zealand, for example:

> The aim is for students to develop a broad technological literacy that will equip them to participate in society as informed citizens and give them access to technology-related careers. They learn practical skills as they develop models, products, and systems. They also learn about technology as a field of human activity, experiencing and/or exploring historical and contemporary examples of technology from a variety of contexts.
>
> (Ministry of Education 2007)

This is a lofty and worthy goal, but what is meant by a broad technological literacy?

Certainly, the study of the historical development of technology and the expectation that pupils will critique their own practice and that of others through the ages is a significant undertaking and a major swing away from the previous craft and vocational aspects of the subject. Some would argue this makes the subject too academic, others that it defines the subject in a modern era: the yin and the yang; very little balance here.

Eastern philosophies hold that yin will eventually turn to yang (and vice versa); this principle pertains to all energy. Yin always gives in to yang and yang to yin; this principle can also be seen in the dualism of D&T education. Vocational gives in to academic and vice versa, depending on the energy of the policy directive. Design and technology education is both an old and a new subject, a juxtaposition that is often not accepted or acknowledged. As an old subject it is associated with developing craft expertise and vocational preparation. As a new subject there is greater emphasis on technology as a general academic subject aimed at developing pupils' understanding in a critical social context. Olsen (1997) described this duality of old and new as the curse of D&T education. The old and the new also represent D&T education as being concerned with different types of knowledge (see Chapter 5 for a discussion of different types of knowledge).

Although many teachers incorporate authentic learning opportunities in their teaching of D&T, making links with pupils' experiences, there is still often a lack of the yin and yang connectedness. The teaching of D&T would be enhanced if it could incorporate the balance of the practical and academic in a connected curriculum. Connecting both facets of the divide is one of the greatest challenges facing D&T education. The challenge requires us to foster pupils' abilities to integrate their learning over a period of time, helping them to develop integrative and metacognitive strategies and habits of mind which will prepare them for the complexities of personal, professional and civic life (Hutchings and Huber 2008; Brears et al. 2011).

Design and technology education has historically been associated with the pressure to meet the demands of society in the vocational arena. This is only one side of the yin and yang symbol and D&T must argue its position in supporting the development of pupils' general knowledge and skills if it is to survive.

Conclusion

My understandings of yin and yang are based on Buddhist adaptations of Taoist philosophy. My understandings of vocational and academic arguments relating to D&T education are based on the works of Dewey (1859–1952). Dewey and the progressive educators challenged the movement that sought to separate academic education for the few and narrow vocational training for the masses. Currently we are faced with the real possibility that they will be separated again and D&T education will be thrust into a state of unbalance.

Dewey promoted the principle of democratic participation, but this can only be achieved through a balance between vocational and academic education and the formulation of an experienced-based connected curriculum. This, I would argue, is the real purpose of D&T education. According to Lakes (1994) radical reproduction theorists have highlighted that vocational education is firmly lodged within the social efficiency tradition, which validates class stratification and perpetuates occupational inequality. A return to separate specialist schooling will undoubtedly fail, as it has in the past, to offer democratic participation. Schools will be used to select and separate teenagers and facilitate the perpetuation of the social classes.

We should not go back to an educational philosophy that belonged to an industrial era in which young people were prepared to be employees and that was their lot. Times have changed, most people do not now spend their working lives in one employment position. We need to help young people to become confident and flexible enough to be able to undertake whatever is required of them. The way to achieve this is through a balanced and broad technological curriculum which does, in fact, develop a modern technological literacy. A balance and harmony of yin and yang, this technological literacy should work alongside other literacies to create enterprising young people who can truly adapt and succeed in the rapidly changing technological society we are facing.

Dewey envisioned participatory democracy in the workplace underpinned by critical pedagogy developed through a balanced education. But Dewey's philosophy has never really been enacted. Developed during the early part of the last century Dewey's vision was perhaps ahead of its time. If D&T education has come of age it has the potential to be the connected curriculum, a modern curriculum that brings balance to the yin and yang of vocational and academic argument by creating a new direction. A combination of liberal education values with the practicalities of new vocationalism, this is the balance that yin and yang design and technology education can bring.

Yin and yang are never static but in a constantly changing balance. When yin or yang are out of balance they affect each other, and too much of one can eventually weaken (consume) the other. I have argued that the overemphasis of one viewpoint of technology education over the other will be unsuccessful. I advocate a balanced curriculum that manages to overturn the overused dualism. If teachers are trying to enhance their technology provision they should consider a complementary yin yang symbol as a positive way forward.

Questions

1. What is your view of design and technology – do you consider it to be an academic or vocational subject?
2. What knowledge and skills do you think are important in a modern technological society?
3. Do you think there is merit in following the philosophy of Dewey?

References

Autio, O. and Hansen, R. (2002) 'Defining and measuring technical thinking: students' technical abilities in Finnish comprehensive schools', *Journal of Design and Technology Education* 14(1): 5–19.

Benn, C. and Chitty, C. (1996) *Thirty years on: is comprehensive education alive and well or struggling to survive?*, London: David Fulton.

Bereiter, C. (2002) *Education and mind in the knowledge age*, Mahwah, NJ: Prentice-Hall.

Brears, L., MacIntyre, W. and O'Sullivan, G. (2011) 'Preparing teachers for the 21st century using PBL as an integrating strategy in science and technology education', *Design and Technology Education: An International Journal* 16(1): 36–46.

Compton, V.J. and Harwood, C.D. (2006) *Discussion document: design ideas for future technology programmes*, http://www.tki.org.nz/r/nzcurriculum/draft-curriculum/technology_e.php (accessed 22 September 2009).

Davies Burns, J. (2000) 'Learning about technology in society: developing a liberating literacy' in Ziman, J. (ed.) *Technological innovation as an evolutionary process*, Cambridge: Cambridge University Press.
Department for Education and Science (1977) Education in schools: a consultative document, Cmnd. 6869, London: HMSO.
Dewey, J. (1948, 1957) *Reconstruction in philosophy*, Boston, MA: Beacon Press.
Dewey, J. (1997) *Experience and education*, New York: Touchstone.
Evans, D. (2008) *A history of technical and commercial examinations: a reflective commentary*, London: TMag.
Evans, R. (undated) *General national vocational qualifications (GNVQs) 1992–2007*, www.technicaleducationmatters.org/node/193 (accessed 29 May 2012).
Gillard, D. (2011) *Education in England: a brief history*, www.educationengland.org.uk/history (accessed 29 August 2011).
Hayward, G., Hodgson, A., Johnson, J., Keep, E., Oancea, A., Pring, R. et al. (2009) *Education for all: the future of education and training for 14–19 year olds*, London: Routledge.
Hutchings, P. and Huber, M. (2008) 'Placing theory in the scholarship of teaching', *Arts and Humanities in Higher Education* 7(3): 229–244.
Lakes, R.D. (1994) 'If vocational education became critical work education', *Journal of Philosophy of Education* 49: 190–195.
Ministry of Education (2007) *The New Zealand curriculum*, Wellington: NZMoE.
Olsen, J. (1997) 'Technology in the school curriculum: the moral dimensions of making things', *Journal of Curriculum Studies* 29(4): 383–390.
Petrina, S. (2000) 'The politics of technological literacy', *International Journal of Technology and Design Education* 10(2): 181–206.
Saunders, M. (2000) 'Understanding education and work, themes and issues' in Moon, B., Peretz, M.B. and Brown, S. (eds) *Routledge international companion to education*, London: Routledge.
School Board of London (1898) 'Report of the Joint Committee on Manual Training on the Development of Work in Connection with Manual Training', 1–24.
Spens, W. (1938) *Secondary education with special reference to grammar schools and technical high schools report of the consultative committee*, London: HMSO.
Tomlinson, M. (2004) *14–19 curriculum and qualifications reform: final report of the Working Group on 14–19 Reform*, London: Department for Education and Skills.
Wolf, A. (2011) *Review of vocational education – the Wolf Report*, London: Department for Education.

Chapter 7

What makes a good technology teacher?

Nigel Zanker and Gwyneth Owen-Jackson

Introduction

> Teaching is a highly complex activity, which requires intellectual sophistication in a dynamic space.
>
> (Howard and Aleman 2008: 161)

There has been much debate, spanning many decades and across many countries, about what constitutes teachers' professional knowledge, what makes teachers effective and how to define 'effective' teaching. This chapter considers these debates and how they might impact on design and technology (D&T) teachers. It also looks at the different routes into teaching and the influence that these might have on the debates.

Teachers' professional knowledge

Across Europe, education has its history in the churches and mosques as it was intended for those entering religious life and for children of the wealthy. It was not until the emergence of industrialisation and urbanisation, in the mid-nineteenth century, that education was made available to the majority. At that time it was easy to define the purpose of education and the role of the teacher, as Dickens (1854) observed, through Gradgrind, all teachers needed to know were 'facts':

> Now what I want is facts. Teach these boys and girls nothing but facts. Facts alone are wanted in life ... nothing else will ever be of service to them ... Stick to Facts sir!
>
> (Dickens 1854)

The teacher had to know his or her subject and then impart it to the pupils, who wrote it down and learnt it. It is a very different story today. As societies have become more complex, and as our understanding of learning has grown, the purpose of education is not so straightforward and the teacher's role is more demanding, as Howard and Aleman's quote at the head of this chapter indicates. There are now shelves of books and journal articles discussing the nature and form of teachers' knowledge, this chapter can do no more than provide a brief summary.

Over 25 years ago Shulman (1987) suggested a typology of seven categories of knowledge needed by teachers and this work still has value today. The categories he identified are:

- content knowledge
- general pedagogical knowledge
- curriculum knowledge
- pedagogical content knowledge
- knowledge of learners and their characteristics
- knowledge of educational contexts
- knowledge of educational ends, purposes and values and their philosophical and historical grounds.

These categories indicate that teachers need to know something about their subject and how to teach it, knowledge of their pupils and schools and knowledge about the value of education.

Building on the work done by Shulman (1987) and others, Leach and Banks (1996) suggested teachers' knowledge comprised subject knowledge, pedagogical knowledge and school knowledge, all informed by an individual's personal constructs (see Figure 7.1). Personal constructs are built on the teacher's own past experience of learning technology, their own belief in the purpose of the subject and their own view of what makes a 'good' teacher. Banks et al. (2006) were keen to stress the 'dynamic relationship' between the different elements in their model, changes in one sphere would have impact on other spheres of knowledge.

Bransford et al. (2005: 11) drew up a similar framework, highlighting teacher knowledge as encompassing:

- knowledge of learners and how they learn and develop within social contexts
- conceptions of curriculum content and goals: an understanding of subject matter
- an understanding of teaching (related to content and learners), informed by assessment and supported by classroom environments.

They also noted that teaching has a moral purpose and 'must serve the purposes of a democracy' (ibid.).

The Training and Development Agency (TDA) in the UK developed a model similar to Leach and Banks as a way of identifying what student teachers needed to learn (see Figure 7.2).

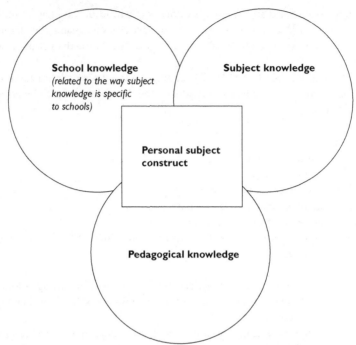

Figure 7.1 A visual tool for describing teachers' professional knowledge (Leach and Banks 1996)

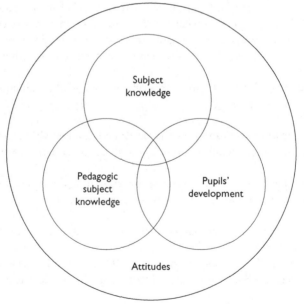

Figure 7.2 Knowledge for Teaching (adapted from TDA 2007)

All these models recognise that teachers' professional knowledge is comprised of their subject knowledge, knowledge of appropriate pedagogy and knowledge and understanding of their pupils and the schools in which they work. However, this list of things that teachers need to know is only a part of their professional knowledge, it is what they do with what they know that makes a difference. As Luntley (2011: 27) points out, 'propositional knowledge does not suffice to capture the knowledge that informs practice'.

As Howard and Aleman (2008) identified, these elements of knowledge that teachers have are constantly interacting in different ways in the classroom and teachers are constantly reconfiguring the knowledge they draw on and use in the classroom. Bransford et al. (2005: 1) noted that teachers 'must be aware of the many ways in which student learning can unfold in the context of development, learning differences, language and cultural influences, and individual temperaments, interest and approaches to learning'. Bondi et al. (2011: 18) recognise that in teaching, and other similar professions, the general knowledge that is acquired then has to be selected for its relevance, appropriateness or sensitivity to each new context; they refer to this as 'professional wisdom'. Teachers draw on their professional wisdom in each new encounter with a pupil or class, in each new lesson. Research has shown that teachers who do this well, those with a deeper knowledge of *how* to teach, as well as *what* to teach, have a positive impact on pupil learning (Hill et al. 2005).

However, this complexity is often not recognised. Shulman's work led, in America, to the introduction of national standards for teacher performance, a move emulated in the UK in 1992 and across other countries. These standards are then used as a way of defining teacher knowledge and a way of measuring it.

Since 1992 the UK government have used standards, or competence statements, to define their expectations of what teachers should know and do, from initial training through to senior positions. The standards, however, tend to be a reflection of the prevailing political agenda rather than an informed consideration of what really constitutes teachers' professional knowledge. The standards in England were updated in 2012 and in these teaching is identified as requiring teachers to have high expectations of pupils, promote progress in learning, have good subject and curriculum knowledge, plan and teach lessons in which they 'impart knowledge', adapt their teaching to suit all pupils, assess pupils, manage pupil behaviour and 'fulfil wider professional responsibilities'. Personal and professional conduct defines the 'behaviour and attitudes' expected of teachers and includes the requirement to 'not undermine fundamental British values' (Department for Education 2011).

So what does this mean for D&T teachers? The requirement to know their subject is discussed later, but what is appropriate pedagogic knowledge in the subject? And how do pupils learn in D&T?

In the UK, as in other countries in Europe, the theory underlying pupil learning has moved from the behaviourist approach through constructivism and social constructivism to situated cognition. There are also theories of constructionism and embodied cognition that are useful to D&T as they assert that there is a strong correlation between doing and thinking. How do teachers use these theories to guide their classroom practice? Do they use these theories?

Teachers' classroom practice is often guided less by their theoretical understanding of how pupils learn and more by their use of curricula, programmes or resources – which may themselves have been developed using learning theory. For example, D&T learning has moved on from the 'dem and do' model, where teachers demonstrated a particular skill and pupils copied and practised it, to engaging pupils in a range of activities. D&T classroom practice across the UK often now includes analysing products, focused tasks in which pupils learn specific skills or knowledge, design and make assignments, working individually and in teams.

Designing in D&T is still often taught in a linear way, whether for ease or convenience or to meet examination requirements is not known. It is, however, widely acknowledged that this is not the way in which designers work, and over 10 years ago Barlex and Welch (2000) showed through their research work that pupils prefer to develop design ideas through 3D modelling rather than sketching. Why then do so many teachers continue to ask pupils to sketch several design ideas, annotate them, choose one, develop and justify it? Are examination requirements driving ineffective classroom practice?

One of the criticisms made of D&T is that it is under-researched (OfSTED 2011). This may be the case, but the criticism can also be made that much of the research work that is being undertaken is not permeating down to impact on classroom practice. And while the research evidence is growing, can it yet be claimed that there is an agreed appropriate, effective pedagogy? And is there sufficient evidence of how pupils learn in D&T to support pedagogy?

Teachers' subject knowledge

One element of teachers' knowledge on which all are agreed is the need to know the subject they teach. How do we define subject knowledge in D&T? Banks (2009: 376), in looking at a number of technology curricula, concludes that 'there is as yet little common agreement as to the subject domain of technology … and to what is valid content'. Should its focus be on technological concepts? If so, which ones, and who decides? Should the focus be on practical skills? (See Chapter 5 for a discussion on this.)

The initial National Curriculum in the UK emphasised the procedural aspects of D&T, as Wakefield and Owen-Jackson point out in Chapter 1, the DES affirmed that 'capability' in D&T processes was as important as 'the acquisition of knowledge' (DES/WO 1989: 1). There were aspects of knowledge listed in the curriculum orders but these were not emphasised in the attainment targets, which specified what pupils should know at different stages of their education. National guidance has continued to focus on the processes of designing and making rather than the knowledge required. Where there is reference to the knowledge pupils should acquire this is often vaguely defined, for example, when studying food pupils' learning should include 'a broad range of practical skills, techniques, equipment and standard recipes' (QCA 2007: 55), which gives no indication of which skills, techniques or equipment they should learn to use.

As Wakefield and Owen-Jackson also highlight (see Chapter 1), when D&T was first introduced into the curriculum, those teaching it were teachers who the previous year had been teachers of art, business studies, CDT and home economics. Those teachers were unsure of the subject knowledge required and the NC provided little guidance. As a result, the subject association for D&T teachers drew up a guidance document identifying the propositional and procedural knowledge that new teachers should be expected to have. Despite criticism from some (Martin 2011) this guidance has undergone two revisions and is still used widely across England and Wales for identifying the required subject knowledge (DATA 1995, 2003, 2010). However, as Ginestie (2009) points out, how subject knowledge is defined for technology teachers varies across Europe.

Although D&T comprises different material areas, UK secondary teachers usually specialise in just one or two. In the early years of the National Curriculum in England there was discussion about D&T teachers being 'generalists', being able to teach in all the areas, but it was soon realised that teachers could not gain the required depth of knowledge across such a broad range of areas.

Primary teachers, however, have to develop subject knowledge across a number of subjects and few of them choose to specialise in D&T. In primary schools, therefore, teachers' D&T subject knowledge is often limited and pupils' experiences constrained. This is not true of all primary schools; there are some in which excellent practice can be found that provides pupils with a sound foundation on which to build, but this is not the case in all primary schools.

One of the difficulties in defining what teachers need to know is that it is not a constant concept, as we can see with our earlier example of Gradgrind. Grant (2008: 129) also identified that 'what teachers need to know, including their skills and dispositions, change and evolve in response to changing social, economic, and political agendas' and Bransford et al. (2005: 3) noted 'that teachers continually construct new knowledge and skills in practice throughout their careers rather than acquiring a finite set of knowledge and skills in their totality before entering the classroom'. This particularly applies to D&T teachers, who have to continually reappraise and update their subject knowledge as new materials and new equipment become available and new processes emerge. Keirl (2007: 45) goes further, saying that 'design and technology's identity is about its capacity to remain dynamic (not a fixed pedagogy or body of knowledge).'

There is also a question to be asked about who decides? Who decides what is appropriate subject knowledge for pupils learning D&T? Is it the government, who prescribe the curriculum? Is it the commercial companies that produce the textbooks and learning resources? Is it the examination bodies that set examination specifications that guide pupil learning? Or is it the teacher and the D&T

educational community? When new technologies and new materials emerge, who decides whether or not these should be on the curriculum for pupils? Curriculum content is not value free, as Briant and Doherty (2012: 53) noted – 'the questions of "which" and "whose" knowledge, skills and values are to be legitimated in any curriculum become a subject of social and political debate'.

Despite the importance given to subject knowledge, and acknowledging that teachers do need to know their subject, there has been some suggestion that this is not necessarily the most important knowledge they need. Research by Goldhaber and Brewer (2000) found that there is not a direct co-relation between a teacher's level of subject knowledge and her effectiveness. Hattie (2012) argues that the amount of subject knowledge a teacher has is less important than how she understands, integrates, organises and presents the knowledge to pupils. A practising teacher also provides some evidence to support this. In an article in *The Guardian* newspaper (19 January 2010) one teacher, recounting experiences from his own school, writes:

> Cameron's [UK Prime Minister] cardinal mistake is to think qualifications make a good teacher. They don't. When you're faced with 30 truculent children after lunch on a Friday afternoon, qualifications don't count for much. Take Lesley, a high-powered business executive who I mentored as she trained to be a teacher. She had everything: a great degree, excellent organisational skills and good communication skills. Yet she crumbled in the classroom because she was so impatient with her pupils, nothing they did was good enough. Whereas her employ had tolerated her endless nit-picking, her pupils became demotivated and disaffected. If you don't have the right personality, you'll suffer in the bearpit of today's classrooms. In my experience there are four types of teacher who are effective: the despot, the carer, the charmer, and the rebel. And none of them, in my experience, requires an upper-class degree.

Teacher effectiveness – so what is a good teacher?

This article identifies, anecdotally, four types of effective teacher. But what do we mean by an 'effective' teacher? Pupils generally judge their teachers on whether they are 'good' or not and their judgements are based on a wide range of factors, including personal relationships, a teacher's enthusiasm, lessons that are interesting and fun, and which involved them (Younger and Warrington 2002). However, in the 'performativity' culture in which schools and teachers now operate, 'effective' teachers are those whose pupils perform well in measures of academic attainment and less attention is now paid to teacher qualities (Stronge et al. 2011).

Bransford et al (2005: 5) acknowledged that 'there is no single "cookie cutter" formula for being successful' but they also go on to say that there are practices and strategies that have been identified as being common to teachers generally regarded as effective in that they are able to support and promote pupil learning. These practices included:

- clearly stating what they expected of pupils
- providing models and exemplars of the work they expected
- moving around all of the classroom while teaching, and monitoring pupils' activities
- many used small group activities
- classroom talk focused on pupils asking questions, sharing ideas and discussing the work
- well-organised lessons with resources prepared and readily accessible.

Many of these elements can be found within the competence statements so presumably all teachers work this way to some degree. So what is it that makes some teachers more effective than others?

Hattie (2009, 2012), after conducting a meta-analysis of hundreds of research reports, found that what makes teachers most effective, that is improve pupil attainment, was good pupil–teacher interactions and teacher 'credibility'. Teacher credibility, according to Hattie, is when pupils believe that a teacher knows the subject, cares about the pupils and wants them to do well; this leads them to trust and respect the teacher and so improves their learning and attainment. This does, of course, mean that teachers need to have well-planned lessons and be well organised but it also requires the ability to be in control of the classroom, good communication skills and confidence. Some of these can be taught and practised, others teachers have to develop for themselves. Hattie also claims that it is important for teachers to be 'passionate and inspired' (2012: 23); where in the lists of competences used across the world is that phrase?

So what would 'effective' D&T teaching look like? The nature of D&T work means that it would not be difficult for teachers to institute the practices described earlier, indeed many already do so. There are many D&T departments in which the learning environment is rich in examples of pupils' work and displays of products and material to provide ideas and inspiration. The best lessons are those that appear effortless but that have been well organised, with the resources ready and accessible, and activities well prepared. The best teachers are those who move purposefully around the room, just happening to be in the right place at the right time, interacting with pupils as they work. It is a joy to be in classrooms like this, when it is clear that both the teacher and the pupils are enjoying the experience.

Preparing effective teachers

Like teaching, teacher education programmes are grounded in their temporal and political context. Programmes vary from country to country, both in structure and in content. This is to be expected, as teachers are prepared to work in the different education systems and, as Banks and Williams show in Chapter 3, for the different forms of technology education that exist.

Despite these differences, most teacher education programmes consist of ensuring that teachers have sufficient subject knowledge, know the accepted pedagogy for teaching that knowledge, how to manage the classroom and know how the school and education system operates.

Currently in the UK, while historically there were a number of routes into teaching, there are now three main routes: higher education institutions (HEI), school based (School Direct) or employment based (School Direct salaried). Higher education routes include undergraduate and postgraduate, with the undergraduate route being the most common for primary teachers and postgraduate the most common for secondary teachers.

Higher education routes include undergraduate courses of three to four years; this allows time for students to study all the subjects required at primary level. Some students choose to specialise in D&T, but these numbers are small and even specialists do not have time for extensive study of the subject. For secondary teachers, the most common route is to take a first degree followed by a one-year postgraduate degree in education (PGCE) with qualified teacher status (QTS). All HEI programmes have close partnerships with schools in which the student teachers undertake placements or practicum, which account for two-thirds of their one-year programme.

School Direct programmes were introduced in the UK in 2012. As the name suggests, they are run by consortia or alliances of schools and only open to graduates. They are available for both primary and secondary teachers, take one year, and lead to qualified teacher status. Most consortia, but not all, also work with local universities and offer PGCE in addition to QTS. School Direct student teachers are based in one school and spend time in a second school, as well as time studying. There is, as yet, little difference between the HEI-based programme and the School Direct programme but as Williams (2009: 535) notes, many countries are moving their teacher education programmes away from the universities and to a more 'school-based apprenticeship model'.

The School Direct salaried scheme differs from School Direct only in that the student teacher is employed, and paid, as an unqualified teacher by the school for the period of training. This teaching requirement means that the programme is aimed at those with at least three years' careers experience, although this does not have to be teaching experience. As with School Direct, student teachers gain qualified teacher status but the option for an academic award (PGCE) may or may not be available.

Teach First is another school-based route into teaching and although numbers are currently relatively small, the government proposes to increase these. Teach First aims to recruit graduates with a minimum 2.1 degree and specific personal qualities. Those accepted onto the programme undergo six weeks of intensive training during the summer, and in September are placed in 'challenging' schools, which are usually those in poor socioeconomic areas, where they begin to teach and continue their training. The programme lasts for a year and provides both QTS and a PGCE. However, Teach First graduates are expected to teach in schools for a minimum of two years and complete 'leadership' training in order to become highly effective classroom teachers and school leaders.

As Maandag et al. (2007) have shown, research into the different models of teacher education is limited and it is difficult to say which produces the most effective teachers. However, it could be argued that the HEI route into teaching gives student teachers a broader understanding of the theoretical knowledge which underpins practice and encourages them to reflect on their own developing practice against this theory. School-based programmes, by the same token, tend to be based on an 'apprenticeship' model in which students observe other teachers and mimic their practice. The debate hinges, therefore, on whether teaching is, as Howard and Aleman (2008) noted, complex and intellectual or whether it is a trade or craft in which students need to learn only a series of procedures.

Conclusion

This chapter has shown that teacher knowledge is multifaceted and much of it is often unacknowledged. Good or effective teachers make the task look simple, but as Howard and Aleman (2008: 161) said 'it requires intellectual sophistication.' This applies even more so in D&T where subject knowledge is fluid and dynamic, the pedagogy develops as subject knowledge and learning theories develop, and teaching has to take account of wider changes in industry, society and the economy. Debate has to continue to discuss what is appropriate subject knowledge, how we can best teach that knowledge and how we can ensure that teachers are effective. But as Bransford et al. (2005: 5) recognised 'Specifying what successful teachers need to know and be able to do is not a simple task.' We would also suggest that it is not a static task, but is dynamic, changing as subject knowledge develops and our understanding of pupil learning and teaching grows. D&T teachers should recognise that they are also lifelong learners.

Questions

1   What elements do you think make up a teacher's professional knowledge?
2   What do you think is important subject knowledge for a D&T teacher?
3   What do you think makes an effective D&T teacher?

References

Banks, F. (2009) 'Research on teaching and learning in technology education' in Jones, A. and de Vries, M. (eds) *International handbook of research and development in technology education*, Rotterdam: Sense.
Banks, F., Barlex, D., Jarvinen, E.M., Owen-Jackson, G., Rutland, M. and Williams, J. (2006) 'Further findings of an International D&T Teacher Education Research

Study: the DEPTH2 Project' in Norman, E.W.L., Spendove, D. and Owen-Jackson, G. (eds) The D&T Association International Research Conference 2006, Wellesbourne, Design and Technology Association.

Barlex, D. and Welch, M. (2000) 'Educational research as a foundation for curriculum development in D&T' in Kimbell, R. (ed.) Design and Technology International Millennium Conference 2000, Wellesbourne, Design and Technology Association.

Bondi, L., Carr, D., Clark, C. and Clegg, C. (2011) *Towards professional wisdom: practical deliberation in the people professions*, Farnham: Ashgate.

Bransford, J., Darling-Hammond, L. and LePage, P. (2005) 'Introduction' in Darling-Hammond, L. and Bransford, J. (eds) *Preparing teachers for a changing world: what teachers should learn and be able to do*, San Francisco: John Wiley & Sons.

Briant, E. and Doherty, C. (2012) 'Teacher educators mediating curricular reform: anticipating the Australian curriculum', *Teaching Education* 23(1): 51–69.

DATA (1995) *Minimum competences for trainees to teach design and technology in secondary schools. Research paper No.4*, Wellesbourne: DATA.

DATA (2003) *Minimum competences for trainees to teach design and technology in secondary schools. Research paper No.4*, Wellesbourne: DATA.

DATA (2010) *Minimum competences for trainees to teach design and technology in secondary schools*, Wellesbourne: DATA.

Department for Education (2011) *Teachers standards effective from September 2012*, London: HMSO.

DES/WO (1989) *Proposals of the Secretary of State for Education and Science and the Secretary of State for Wales. Design and Technology for Ages 5 to 16*, London: HMSO.

Dickens, C. (1854) *Hard Times*.

Ginestie, J. (2009) 'Training technology teachers in Europe' in Jones, A. and de Vries, M. (eds) *International handbook of research and development in technology education*, Rotterdam: Sense.

Goldhaber, D. and Brewer, D. (2000) 'Does teacher certification matter? High school teacher certification status and student achievement', *Educational Evaluation and Policy Analysis* 22(2): 129–145.

Grant, C. A. (2008) 'Teacher capacity' in Cochran-Smith, M., Feiman-Nemser, S., McIntyre, D.J. and Kemers, K.E. (eds) *Handbook of research on teacher education*, New York: Routledge.

Hattie, J. (2009) *Visible learning: a synthesis of over 800 meta-analyses on achievement*, London: Routledge.

Hattie, J. (2012) *Visible learning for teachers maximizing impact on learning*, Abingdon: Routledge.

Hill, H., Rowan, B. and Ball, D. (2005) 'Effects of teachers' mathematical knowledge for teaching on student achievement', *American Educational Research Journal* 42(2): 371–406.

Howard, T.C. and Aleman, G.R. (2008) 'Teacher capacity for diverse learners. What do teachers need to know?' in Cochran-Smith, M., Feiman-Nemser, S., McIntyre, D.J. and Kemers, K.E. (eds) *Handbook of research on teacher education*, New York: Routledge.

Keirl, S. (2007) 'Discomforting the orthodox: using debates in the pedagogy of curriculum and critical thinking in design and technology teacher education' in

Norman, E.W.L. and Spendlove, D. (eds) The D&T Association Annual Research Conference 2007, Wellesbourne, Design and Technology Association.

Leach, J. and Banks, F. (1996) *Investigating the developing 'teacher professional knowledge' of student teachers*. Paper presented at BERA Conference, Lancaster, September.

Luntley, M. (2011) 'Expertise – initiation into learning, not knowing' in Bondi, L., Carr, D., Clark, C. and Clegg, C. (2011) *Towards professional wisdom: practical deliberation in the people professions*, Farnham: Ashgate.

Maandag, D.W., Folkert Deinum, J., Hofman, W.H.A. and Buitink, J. (2007) 'Teacher education in schools: an international comparison', *European Journal of Teacher Education* 30(2): 151–173.

Martin, M. (2011) 'From facts to fractals: exploring theoretical perspectives on (subject) knowledge for design and technology' in Stables, K., Benson, C. and de Vries, M. (eds) *PATT25: CRIPT8 perspectives on learning in design & technology education*, London: Goldsmiths University of London.

OfSTED (2011) *Meeting technological challenges? Design and technology in schools 2007–10*, London: HMSO.

QCA (2007) *Design and technology programme of study for Key Stage 3 and attainment target*, London: HMSO.

Shulman, L.S. (1987). 'Knowledge and teaching: foundations of the new reform', *Harvard Educational Review* 57(1): 1–22.

Stronge, J.H., Ward, T.J. and Grant L.W. (2011) 'What makes good teachers good? A cross-case analysis of the connection between teacher effectiveness and student achievement', *Journal of Teacher Education* 62(4): 339–355.

TDA (2007) *Developing trainees' subject knowledge for teaching*, London: HMSO.

Williams, P.J. (2009) 'Teacher education' in Jones, A. and de Vries, M. (eds) *International handbook of research and development in technology education*, Rotterdam: Sense.

Younger, M. and Warrington, M. (2002) 'He's such a nice man, but he's so boring, you have to really make a conscious effort to learn': the views of Gemma, Daniel and their contemporaries on teacher quality and effectiveness' in Moon, B., Shelton Mayes, A. and Hutchinson, S. (eds) *Teaching, learning and the curriculum in secondary schools: a reader*, London: RoutledgeFalmer.

Part III

Debates within design and technology

Chapter 8

Does food fit in design and technology?

Suzanne Lawson

Introduction

Yes, it does. No, it does not. The debate here is not issues within the subject; the debate is the subject itself. Some argue that design and technology (D&T) is not where food studies should be located in the school curriculum, others argue that it is. This chapter will look at the debate around food within the D&T curriculum and will consider the historical context, gender issues, food and health, vocational education and practical issues. It is hoped to offer a range of opinions so that you can draw your own conclusions and decide – does food fit in design and technology?

The debate is focused on food technology in the secondary school as in primary schools, although food technology is taught as part of the D&T curriculum, it is often taught within a cross-curricular project. Teaching food in this way allows for natural links to be made with science, mathematics, health and well-being and there seems to be less concern over whether or not it should form part of D&T (Rutland and Miles-Pearson 2009).

Historical context

Food as a subject in the English curriculum is well established: British Museum records trace cookery schools in London to 1740, and food in the school curriculum dates back to the 1840s (Rutland 2006). Food education at this point was practical, philanthropic and utilitarian, providing knowledge and training for the lower classes, mainly teaching cookery skills to girls so that they could feed families and prepare for domestic auxiliary occupations.

Since the beginning of the twentieth century a number of educational reports have reinforced the teaching of food studies as a practical subject concerned with preparing nutritious family meals. In 1926 the Howden Report (Central Advisory Council for Education (CACE) 1926: 235) stated: '[I]n cookery the general aim should be to provide practical instruction in the choice and preparation of the food required for a simple wholesome diet, with due regards to home conditions and the need for economy.' For the middle classes, selective grammar schools taught girls domestic science with an emphasis on nutrition and science (Rutland 2006). In 1945 the new secondary modern schools adopted a rationale that girls should be taught how to cook to restore the health of the nation after a period of austerity (Rutland 2006). Although the Crowther Report (CACE 1959) considered the changing social and industrial needs in society, post-war food education was still grounded in housecraft traditions for girls. The Newsome Report (CACE 1963) documented the subject's ability to offer creative and civilising experiences beneficial to all pupils but still differentiated by gender, 'housecraft and needlework easily justify their place in the curriculum for most girls' (p. 389). It could be argued that such food education sat comfortably in the social climate of the times with reports such as these supporting its philanthropic–utilitarian function.

By the mid-twentieth century domestic science had evolved into home economics, although it was still focused on preparing meals for families (DES 1985) and still studied mainly by girls. In contrast, craft, design and technology (CDT) (DES 1987) was studied mainly by boys and, through various schemes and initiatives introduced into schools, had begun to focus on problem solving and producing creative practical solutions to design problems.

Towards the end of the twentieth century, the introduction of technology in the National Curriculum (DES 1990), and the inclusion of home economics in this new subject, (see Chapter 1) was a turning point in the subject's evolution and marks the beginning of this debate. Its inclusion in technology meant that home economics teachers became 'food technology' teachers and the emphasis moved from domestic production of meals to commercial production of food products. Food teachers received no training and little support for teaching the industrial and commercial aspects of food production and many felt alienated from a curriculum that seemed to have little reference to food and contained words such as designing, artefacts, systems and mechanisms (Rutland 2006). The content of the subject changed with the introduction of vocabulary such as product development, sensory evaluation, food for a target market of consumers and increasing use of industrial tools and equipment. Technology married together traditional craft subjects but appeared to be closer to the philosophy of CDT than home economics. Did food, with its history of domestic study, fit this philosophy? Did it have anything in common with technological areas? While some teachers felt alienated by these changes in the subject, others saw the transformation as presenting new and exciting challenges, opportunities to move the teaching forward and the best opportunity for the future of food in schools. Is this still true?

Critics of the National Curriculum, most notably from the engineering community, do not think food belongs in the D&T curriculum. Their arguments include that being able to cook is 'affected by technology but [is] not necessarily part of it' (Smithers and Robinson 1992: 15) and that the inclusion of food in technology was more about keeping home economics alive and making technology girl friendly than on its intrinsic value (Smithers 1993, cited in Fine 1994).

The Engineering Council, which continues to be vocal about the content of D&T, did concede that food had a place in the curriculum, but not within technology. As Fine (1994: 41) notes: 'The gist of the anti-food lobby was that Technology was essentially about making, structures and artefacts so to include food would reduce its rigour and diminish its status.' In contrast, the British Nutrition Foundation (BNF) welcomed the new curriculum, noting that food technology should not be confused with cookery and arguing that food technology teaches pupils scientific principles (Fine 1994). Understanding and communicating this argument is central to the debate about whether or not food fits in D&T.

Revision of the National Curriculum in 1995, and the change in nomenclature from technology to design and technology (DFE/WO 1995), did nothing to help. Teachers wanting to teach food skills in a traditional context found themselves teaching pupils the, arguably worthless, activity of drawing food products and requiring them to complete paper-based 'designing tasks' rather than providing opportunities to develop practical skills. Pressures on curriculum time for D&T meant that many schools adopted a 'carousel' approach, in which pupils spent just a few weeks each year in each of the D&T areas, resulting in limited opportunities in which to cook, which were then further restricted by demands to draw food products and packaging. Food teachers struggled with the conflicting demands of developing pupils' practical food skills as well as their design capabilities. Criticism from OfSTED (2002, 2004, 2005) highlighted that some teachers paid insufficient attention to the process of designing, limiting pupils' experience to a sequence of short, focused practical tasks. Without explaining why this was important in food technology, the OfSTED reports reinforced the idea that pupils should draw their food design ideas, leaving many food teachers frustrated. The Key Stage 3 national strategy 'Design and technology framework and training materials' (DfES 2004) (government-produced materials intended to support teachers in the teaching of designing) provided a broader definition of the term 'designing' and offered teachers non-graphical examples of designing and modelling strategies, which could be used in food technology, such as using the 'matrix approach', 'product pairs', 'word association' and 'extending the product range'. This was helpful, but it still presented design work as a paper-based activity.

What seems to have been lost at the time was the realisation that designing in food was not necessarily about drawing. A student interviewed as part of Rutland and Barlex's research concluded that designing in food should be a concurrent activity in which pupils handle food, learn new skills and knowledge and develop ideas. Rutland and Barlex (2006) argued that designing in food was different from other areas of D&T as it was a 'simultaneous activity' (p. 6) involving brainstorming, questionnaires, attribute and product analysis as well as modelling with food. This view is supported by Owen-Jackson (2007) who argues that designing in food is better referred to as 'food product development' and involves working with the food ingredients rather than drawing food products.

Further revisions to the National Curriculum in 1995 (DFE), 1999 (DfEE/QCA) and 2007 (QCA) all retained food within D&T. While the arguments abated somewhat there was still an underlying tension as to whether or not food fitted with the D&T philosophy and curriculum. There was also some polarisation in the teaching of food technology. Those teachers who embraced it made the subject interesting, challenging and motivating for pupils; those who didn't found that they could teach the Licence to Cook programme to younger pupils and catering examinations to older ones. This meant that some pupils were receiving a science- and industry-based food technology learning experience, others were learning practical skills. As the National Curriculum in England is about to undergo further review it will be interesting to see where food is located and the philosophical approach taken to its content.

Gender issues

The historical development of food in the school curriculum describes how it was focused on educating girls (Attar 1990). This gender bias continued through to 1975, when the Sex Education Act in the UK made it illegal to discriminate on grounds of gender in many aspects of social life, including schooling and the curriculum. This meant that schools had to allow both sexes to study all subjects on the curriculum (Rutland 2006), although legislating for this to happen did not necessarily mean that it would.

Following equal opportunity reforms the number of boys taking home economics at examination level increased from one in 10 pupils in 1975 (when the Sex Discrimination Act was introduced) to one in nine in 1985 (Attar 1990: 21). Attar pessimistically predicted that it would be well into the twenty-first century before the subject recruited equal numbers of boys and girls and examination entry data seem to support this. Table 8.1 shows the number of boys and girls taking home economics as an examination at age 16 and how the percentage of boys has remained low.

Food technology has fared better. Table 8.2 shows the number of boys and girls taking it as an examination at age 16. Although the number of boys entering is steadily increasing each year, the 50% target is still some way off.

Table 8.1 Home economics examination entries classified by gender (000s)

Year	Boys	Girls
2010	3.1 (11%)	25.6
2009	2.9 (10%)	27.6
2008	3.0 (9%)	31.9
2007	2.5 (7%)	34.1
2006	2.3 (6%)	33.2

(Source: www.data.org)

Table 8.2 Food technology examination entries classified by gender (000s)

Year	Boys	Girls
2010	22.4 (36%)	39.7
2009	23.5 (35%)	43.2
2008	25.1 (34%)	49.1
2007	25.1 (33%)	55.0
2006	25.9 (30%)	59.3

(Source: www.data.org)

These figures seem to suggest that the industrial emphasis of food technology courses is more attractive to boys than the traditional approach of home economics. The Office for Standards in Education (OfSTED) (2011) concludes that boys select food courses if they are practical and engaging, suggesting perhaps that they appeal to boys interested in careers in catering or the food industry. Rutland (2006) sees the focus on job opportunities offered by food technology as beneficial (see Chapter 6 for a discussion on the vocational versus academic debate). Other ways of encouraging more boys to study food include encouraging more men to become food teachers and providing role models, marketing materials to counter traditional stereotypes and highlighting the contribution food technology courses can make towards future economic well-being, careers and lifestyles (OfSTED 2011).

Research conducted by Davies et al. (2008) looked at socioeconomic background, gender and subject choice in schooling using data from 112,412 pupils. Their research found that not only gender but also socioeconomic background was influential in pupils' choice of examination subjects. Pupils from higher status socioeconomic backgrounds were more likely to be entered for examinations in academic subjects such as French or history than in 'applied subjects' such as home economics. This indicates that there is a need to challenge perceptions about food education; there seems to be a lack of understanding among policymakers and parents, head teachers and pupils about what food education teaches. Home economics was concerned not only with practical skills but also with the scientific concepts of nutrition, the chemical and nutritional properties of foods, heat transfer, the effects of heat on food and the practical application of these concepts.

Similarly, food technology is not only about cooking, following a recipe or making a product (Rutland 2006) but is underpinned by the key concepts of design and making, cultural understanding, creativity and critical evaluation. Pupils still learn the scientific concepts identified earlier but they also learn how to design/develop food products, model and communicate their ideas, respond creatively to design briefs; they develop proposals and specifications, applying their knowledge and understanding of ingredients, make a range of products using different ingredients and skills and are constantly evaluating what they do.

Critically, they solve technical problems using hand and machine tools, including computer-aided design and manufacturing techniques (CAD/CAM). Pupils learn a broad range of practical skills and techniques, they learn about safety and hygiene, healthy eating, balanced diets, nutritional needs and the nutritional, functional and sensory properties of food. Learning about food also makes a valuable contribution to pupils' developing understanding of cultural issues. OfSTED (2011) commended how culture could be addressed in a meaningful manner when designing and making in food. Pupils learn about the qualities of different ingredients and their significance in different cultures, as well as why and how they reflect associated cuisines. Food technology is not about rote learning of skills but requires 'thought in action' (Rutland 2006: 277). Part of design and technology's uniqueness in the curriculum is the demand for pupils to investigate, explore possible solutions, modify, evaluate and develop ideas.

It can also be argued that many aspects of food technology are related to science and there is evidence that girls and women are still under-represented in this area (see Smith 2011). Food technology is one way of interesting and motivating girls' (and boys') interest in scientific issues and increasing girls' participation in science-related learning.

This is not to say, of course, that good practice is evident in all schools. Bielby (2005) and OfSTED (2006) were critical of courses that over-emphasised large-scale food production methods at the expense of traditional skills. OfSTED (2006: 10) articulated the concern that 'too little time is spent learning to cook nutritious meals and too much time is devoted to low level investigations and written work'. In contrast, there are some teachers who focus too much on pupils' learning practical skills at the expense of developing their broader understanding of food ingredients. There is also some concern that the food technology curriculum needs to be modernised to make it appropriate for pupils in the twenty-first century (Rutland and Owen-Jackson 2012).

Food technology, then, suffers from being a subject perceived as being 'for girls' and as a subject with more vocational relevance than academic. Those who argue that food does not fit in D&T tend to hold these perceptions. However, if food technology is perceived as a subject that teaches scientific concepts, allows pupils to design and be creative with food through modelling and developing, learn about social and cultural values, solve problems and meet social needs then, it can be argued, it does fit within D&T.

Food and health

Learning about food at all ages, where it comes from, what we should eat and how we can prepare it is a crucial aspect of combating a range of health issues including heart disease, strokes, cancer, diabetes and obesity (Rutland and Barlex 2009). Undoubtedly, social changes and lifestyle issues have impacted on home cooking; time spent on food preparation is now 20 to 40 minutes per day compared to the two hours of 50 years ago (Popkin 2008) and there is less use of fresh ingredients. Food tends to be available at different times for different family members, which often means fewer opportunities for young people to be involved in cooking (Popkin 2008). Horne and Kerr (2003) found, when investigating the lifestyles of 1,200 Scottish school children, that education ministers had made an erroneous assumption that schoolchildren were taught how to cook at home. Many believe that this can be combated through improving food education in school: '[I]t [healthy eating] has to start with education' (Kelley et al. 2011: 41) and that schools must play an increasing role in obesity management and prevention.

Concerns about rising obesity rates and media campaigns about children's eating habits led the UK government to consider the place of food education on the school curriculum. In 1999 it introduced the 'Healthy Schools' initiative, a non-statutory programme in which 97 per cent of schools participate (www.idea.gov.uk) and food technology can make an important contribution to this (Owen-Jackson 2007). In 2008 the government report *Healthy Weight, Healthy Lives* (DH/DCSF 2008) introduced the 'Licence to Cook' programme into schools. Licence to Cook provided all pupils aged 11–14 years with an entitlement to eight hours each year to learn to cook nutritious dishes from basic ingredients, whether or not their school offered cooking as part of the curriculum. The programme was intended to teach basic recipes, the principles of diet and nutrition, health and safety and wise food shopping. Teachers were trained to teach the programme and use the extensive range of resources developed by the British Nutrition Foundation (www.licencetocook.org.uk). This in itself is interesting; if the teachers of Licence to Cook were already trained home economics/food technology teachers then why would they need further training in order to teach this programme? If the teachers were not food teachers, then it is likely that the training provided only rote learning of the recipes and practical skills involved, which is what the pupils would have been taught. Many food technology teachers regarded this as an impoverished learning experience.

However, despite the reservations of some teachers, 3,000 secondary schools registered for the programme and the emphasis in food education moved to basic practical skill development. The eight hours each year taken by the Licence to Cook programme were, in many schools, taken from the time allocation for food technology meaning that it was being taught in place of food technology. OfSTED (2011) acknowledged the emphasis on cooking and healthy eating in schools during 2007–10, noting how the Licence to Cook initiative led to more opportunities for practical work and a greater concentration on making meals. For many food technology teachers this was a welcome return to the philanthropic–utilitarian function of food education. Others such as Rutland (2008) saw the move away from designing and making as compromising food technology at Key Stage 3. In her article 'Licence to Cook: the death knell for food technology', Rutland (2008) critiques the Licence to Cook programme, identifying potential problems related to curriculum time, resources, progression from Key Stage 3 to 4 and compliance with the D&T ethos. She notes that food education is far more than a craft approach to practical cookery skill development. Within D&T food technology has a much wider remit that focuses on user needs. Through developing skills in research, analysis, synthesis and evaluation pupils solve problems using ingredients, equipment and technology.

The Licence to Cook programme was not the only response to the government's concerns about the nation's health. Other schemes included 'Jamie's School Dinners', a campaign led by Jamie Oliver to improve school dinners, the School Fruit and Vegetable Scheme, the Every Child Matters agenda, the Big Lottery-funded Let's Get Cooking campaign and Healthy Schools. Rutland and Barlex (2009: 17) report that the Foresight Project concluded that over recent years the United Kingdom has become a nation 'where overweight has become usual rather than unusual'. In 2008 the cross-government Obesity Unit report (DH/DCSF 2008) included aspirations to improve healthy eating, physical activity and the need to make food technology compulsory in the National Curriculum.

The health issues which have raised government concerns highlight the importance of food in the school curriculum. This could, of course, suggest that learning practical skills is important and that food studies should, therefore, not be part of D&T. However, learning these skills without understanding the scientific concepts that underpin cooking and nutrition is less likely to impact on pupils' behaviour. This argument still seems to suggest that food should fit within the D&T curriculum if pupils are to properly understand the role of food in relation to health.

Food education as vocational education

Food education has long-standing associations with vocational education. In 1926 the Howden Report (CACE 1926: 234) considered that courses for housecraft, which included cookery, should be planned so as 'to render girls fit on leaving school to undertake intelligently the various household duties which devolve on women'. Today the food industry is a large and influential body with opportunities for high-status careers in a wide range of roles providing food with a vocational purpose on the curriculum. It could be argued that the inclusion of food within D&T helped to bring the subject up to date by situating the learning within an industrial context. Bielby's research (2005) found that pupils gained an understanding of the wide range of jobs within the food industry and boys seemed to enjoy the subject more when there was less domestic influence and it was not seen as a 'girls' subject'.

However, teaching about industrial practice can be a challenge for food teachers. Many food technology teachers have little industrial experience and have had to update their own knowledge and skills using commercially produced teaching materials and through friends and acquaintances in the food industry (Bielby 2005). There are also challenges in providing pupils with authentic industrial experiences. Industry standard equipment is expensive so schools buy one piece of equipment for all pupils to use, or scaled-down versions of equipment, or pupils only see industrial equipment being used on video or DVD programmes. Initially, teachers tried to arrange visits to manufacturing sites but, even where manufacturers were supportive, as demand increased it became more difficult to provide such an experience.

With more interest in vocational courses in schools there has been an increase in the availability and uptake of catering courses. Many teachers feel more comfortable with teaching catering and it is certainly easier to provide authentic learning experiences for pupils.

The link between food and vocational interests seems to suggest that food does not fit with D&T, that it should be available on the school curriculum in its own right as a vocational subject. While there is some merit in this argument, such a move would mean that food studies would only become available to pupils at age 14 and there is a need for their food education to begin much earlier than this. It is also likely that the focus of food education, were it to become vocational, would be on the practical skills needed for catering. The food manufacturing industry is much greater than this and requires graduates with a wide knowledge base and a broader range of skills, such as is provided by food technology (Monks 2012). Again, this suggests that the best place for the teaching of food education is within D&T where it is able to teach knowledge and skills beyond the boundaries of 'practical work'.

Practical issues

Finally, it has to be acknowledged that D&T is an expensive subject for schools because class sizes are usually smaller, the tools and equipment are expensive and there is a constant requirement for materials and ingredients. Some, therefore, question whether schools can afford to teach D&T, including food technology, and whether its educational benefit is worthwhile.

The practical issues in relation to food technology are to do with cost, time and class size. As with the other D&T areas, there are costs associated with supplying and maintaining equipment, updating equipment and tools and providing chilled storage facilities. In addition, unlike most other subjects on the school curriculum, food technology usually requires pupils to provide their own ingredients for lessons. There are some schools, particularly those in areas of high economic and social need, that fully or partially fund food ingredients (OfSTED 2011) but this is not the norm. Requiring pupils to provide ingredients can impact on their 'active inclusion' (OfSTED 2011: 42) in lessons or can constrain what they learn (Owen-Jackson 2007). If schools were able to provide ingredients there would be more scope for pupils to engage in experimental and investigative work and to be more creative in their food 'designing' work. Economic constraints mean that this is unlikely to happen, but if it were it would provide pupils with a richer learning experience.

The curriculum time for teaching food technology also imposes constraints on what can be taught. Most lessons are 45 to 50 minutes long, which includes time for pupils to arrive and prepare for the lesson and to clear and pack away at the end. This time limit means that much of pupils' practical work is limited to biscuits, scones, pizza, sauces, dishes that can be quickly prepared and cooked. In schools in which lessons are longer pupils are able not only to undertake more challenging practical work but also to spend time on evaluating their work and that of their peers, again providing a more enriching learning experience.

Class sizes can also prove challenging (Elliott 2009). Food technology class sizes are often determined by senior managers in the school who are not aware of the limitations imposed by the availability of equipment and the demands and potential dangers of practical activities. Large class sizes often result in pupils' limited use of equipment and tools or constraints on their practical work based on what is feasible rather than what would be a good learning experience.

These practical issues can be cited as reasons for not including food technology in D&T, but they apply equally across other areas of the subject and should not be considered in isolation. Food technology, as other areas of D&T, provides pupils with a worthwhile educational experience that develops their knowledge and skills in a way that no other subject does. It allows pupils to work conceptually and practically; they learn not only about food ingredients but also about underlying scientific concepts, they learn to define and understand problems and look for feasible solutions and, in doing so, develop a wide range of valuable skills.

Food technology or food studies?

Prior to the introduction of the National Curriculum in the UK, food studies had its place on the curriculum as home economics taught to girls and boys as a life skills course and vocational preparation. The subject was not highly valued but most pupils experienced and enjoyed it. Since the inception of the National Curriculum, food technology has been part of the D&T curriculum and some have argued that this is not the place for food education. Assuming that there is still a place for food studies on the school curriculum, if it is not going to be within D&T where will it be?

One proposal is that it could be taught through personal, social and health education (PSHE). This would likely see a return to food education focused on families, healthy eating, diet and health and personal well-being (QCA 2007); it would also mean the loss of practical food preparation skills. Rutland (2006) suggests that food education within PSHE would mean that it would be taught by non-specialists not committed to food education, it may lack co-ordination and become a fragmented experience and, as a result, it would have low status.

An alternative is that food education as a separate subject could be removed from the curriculum, as other subjects teach much of the content. Science, for example, teaches about growth, development, behaviour and health in relation to diet and the chemical and nutritional aspects of food; geography teaches human issues, including sustainability in relation to food production; citizenship considers the legal right of consumers and the changing nature of UK society including food. Surprisingly, physical education (PE) makes little reference to food. With this proposal, food education would become cross-curricular and there would be no co-ordination to ensure that pupils could see the 'bigger picture' in relation to food. Also, these subjects already have crowded curricula; if food topics were added then little time would be available to teach them thoroughly. In addition, within science the food concepts would be taught in an abstract way and many pupils find abstract knowledge difficult to understand. In a food technology context, these concepts would be taught through concrete learning and would be applied, developing both pupils' understanding and their higher order thinking skills such as research, analysis, synthesis, decision making, time management, organisation, evaluation, critical thinking and, not least, problem solving to meet the needs of a user.

OfSTED (2006: 1) identified that one of the problems for food technology is that 'there is a fundamental and so far unresolved dichotomy between teaching about food to develop skills for living and using food as a means to teach the objectives of D&T.' There are those who still see food education as being a 'life skill', requiring pupils only to learn how to cook healthy and nutritious meals in order to feed themselves and their families. Others see food education as broader than this, encompassing scientific concepts that are taught in a meaningful way to pupils.

Conclusion

If food education is seen as having a narrow 'craft skill' and a utilitarian function it may fit into the curriculum but not into D&T. Many food educators would argue, and this chapter has shown, that the subject can offer more than just these skills.

The changes in school-based food education in the last 20 years have meant inevitable pedagogical development. Home economics had a strong focus on pupils learning specific skills, a focus that was lost during the formative years of the National Curriculum. In the early years of the National Curriculum food education struggled to fit within D&T, mainly due to the formulaic and ritualistic interpretations of the requirements leading to pupils being asked to draw food design solutions. This is less in evidence now and teachers are beginning to develop their own, meaningful pedagogy for food technology.

Within D&T food education provides an opportunity to think and investigate in practical ways by engaging with ingredients, processes, equipment and technologies to find solutions for a user and purpose. The subject allows pupils to apply their scientific and mathematical understanding to solve problems. In doing this, it can also contribute to educating pupils about their health and well-being. And if the purpose of food education is also to provide a vocational context leading to employment then the creative, technical, analytical, evaluative and interpersonal skill development opportunities will prepare pupils for a wide range of career opportunities within the food industry, some of them at a high level.

It has to be acknowledged that there may be a case for food technology to be updated, despite the challenges of doing so (OfSTED 2011). Pupils are entitled to learn about up-to-date scientific and technological research, developments such as functional foods and synthetic flavours and 'innovative new materials and to investigate practically how and why they work' (OfSTED 2011: 4). There is a need to make the subject more relevant and more challenging (Rutland and Owen-Jackson 2012), but this is not a reason for excluding it from D&T. Policymakers, parents, head teachers and pupils need to be made aware that a modern, updated food technology curriculum is a worthwhile part of every pupils' general education.

Has food education over the years done nothing more than 'systematically waste girls' time'? (Attar 1990). Perhaps in earlier years that was a valid concern, but as Rutland (2008: 67) asks: '[W]ould a return exclusively to teaching meal planning, including the needs of individuals in the family, invalid cookery, vegetarian cookery, marketing and table-laying be appropriate for pupils in the twenty-first century?' The obvious answer to this is 'no' and food technology does not supply this. The debate should not be whether or not food fits into the D&T curriculum; rather it should be how can we make food technology more relevant and challenging for pupils in the twenty-first century?

Questions

1   What should be the main focus of a food technology curriculum – is it a practical-based subject or a science-based subject?
2   If you could design your own food education curriculum what would it contain? What would be the key words and how would it reflect the learner in the twenty-first century?
3   What is the role of food within the D&T curriculum?

References

Attar, D. (1990) *Wasting girls' time – the history and politics of home economics*, London: Virgo.
Bielby, G. (2005) 'Teachers' experiences of teaching young people about the food industry' in Norman, E. and Spendlove, D. (eds) The Design and Technology International Research Conference, Wellesbourne: Design and Technology Association.
Central Advisory Council for Education (1926) *The education of the adolescent. (The Howden Report)*. London: HMSO.
Central Advisory Council for Education (1959) *15–18. The Crowther Report*. London: HMSO.
Central Advisory Council for Education (1963) *Half our futures. (The Newsome Report)*. London: HMSO.

Davies, P., Telhaj, S., Hutton, D., Adnett, N. and Coe, R. (2008) 'Socioeconomic background, gender and subject choice in secondary schooling', *Educational Research* 50(3): 235–248.

Department for Education and Science (DES) (1985) *Home economics from 5–16 – curriculum matters*, London: HMSO.

DES (1987) *Craft, design and technology from 5–16*. London: HMSO.

DES (1990) *Technology in the National Curriculum*, London: HMSO.

Department for Education (DFE)/Welsh Office (WO) (1995) *Design and technology in the National Curriculum*, London: HMSO.

Department for Education and Employment (DfEE)/Qualifications and Curriculum Authority (QCA) (1999) *The National Curriculum*, London: Author.

Department for Education and Science (DES) (1985) *Home economics from 5 to 16 – curriculum matters*, London: HMSO.

Department for Health/Department for Children, Schools and Families (DH/DCSF) (2008) *Healthy weight, healthy lives*, London: Department of Health Publications.

DfES (2004) *Key Stage 3 national strategy: design and technology framework and training materials*, London: HMSO.

Elliott, G. (2009) 'Issues associated with teaching practical cookery in UK schools: evidence from a survey of teaching staff' in Norman, E. and Spendlove, D. (eds) The Design and Technology Association International Research Conference Proceedings, Wellesbourne: Design and Technology Association.

Fine, G. (1994) 'Is there a future for food education in schools?', *Design & Technology Teaching* 26(3): 39–44.

Horne, S. and Kerr, K. (2003) 'Equipping youth for the 21st century. The application of TOWS analysis to a school subject, *Journal of Non-profit and Public Sector Marketing* 11 (2).

Kelley, T., Carroll, S., Jones, G. and White, J. (2011) *Obesity epidemic – paranoia or evidence based?*, The Bow Group Health and Education Policy Committee, London: Author.

Monks, B. (2012) 'Chilled education investing in the future', *D&T Practice* 1: 2012.

OfSTED (2002) *Report 2000/01*, London: HMSO.

OfSTED (2004) *OfSTED subject reports 2002/03 – design and technology in secondary schools*, London: HMSO.

OfSTED (2005) *OfSTED subject reports 2003/04 – design and technology in secondary schools*, London: HMSO.

OfSTED (2006) *Food technology in secondary schools*, London: HMSO.

OfSTED (2011) *Meeting technological challenges? Design and technology in schools 2007–10*. London: HMSO.

Owen-Jackson G. (ed.) (2007) *Practical guide to teaching design and technology in the secondary school*, London: Routledge.

Popkin, B. (2008) *The world is fat: the fads, trends, policies, and products that are fattening the human race*, New York: Avery.

QCA (2007) *The National Curriculum Key Stage 3*, London: Author.

Rutland, M. (2006) 'The inclusion of food technology as an aspect of technology in the English school curriculum: a critical review', in de Vries, M.J. and Mottier, I. (eds) *International handbook of technology education – reviewing the past twenty years*, Rotterdam: Sense Publishers.

Rutland, M. (2008) 'Licence to Cook: the death knell for food technology?', The Design and Technology Association International Research Conference Proceedings, Wellesbourne: Design and Technology Association.

Rutland, M. and Barlex, D. (2006) 'Developing a conceptual framework for auditing design decisions in food technology: the potential impact on initial teacher education (ITE) and classroom practice' in Norman, E. and Spendlove, D. (eds) The Design and Technology International Research Conference Proceedings, Wellesbourne, : Design and Technology Association.

Rutland, M. and Barlex, D. (2009) 'The politics of food: inside and outside of school' in Norman, E. and Spendlove, D. (eds) The Design and Technology Association International Research Conference Proceedings, Wellesbourne: Design and Technology Association.

Rutland, M. and Miles-Pearson, S. (2009) 'The position of food in the primary school curriculum: implications of the review of the primary curriculum in England' in Benson, C., Bailey, P., Lawson, S., Lunt, J. and Till, W. (eds) Proceedings of the 7th International Primary Design and Technology Conference, Birmingham: CRIPT.

Rutland, M. and Owen-Jackson, G. (2012) *Current classroom practice in the teaching of food technology: is it fit for purpose in the 21st century?* Paper presented at PATT26, Stockholm, June 2012.

Smith, E. (2011) 'Women into science and engineering? Gendered participation in higher education STEM subjects', *British Educational Research Journal* 37(6): 993–1014.

Smithers, A. and Robinson, P. (1992) *Technology in the National Curriculum: getting it right*, London: Engineering Council.

Further reading

Barlex, D. and Rutland, M. (2008) 'Justifying the inclusion of design and technology in the school curriculum. A small-scale investigation into the views of those about to enter the teaching profession concerning reasons for teaching design and technology', PATT20 Conference Proceedings; www.iteaconnect.org/Conference/pattproceedings.htm

OfSTED (2008) *Education for a technologically advanced nation*, London: HMSO.

Rutland, M. (2010) 'Food technology in secondary schools in England: its place in the education of a technologically advanced nation' in Norman, E. and Spendlove, D. (eds) The Design and Technology Association International Research Conference Proceedings, Wellesbourne: Design and Technology Association.

Valentine, S. (2009) 'Healthy eating – never mind the policy – this is what we teach!' in Norman, E. and Spendlove, D. (eds) The Design and Technology Association International Research Conference, Wellesbourne: Design and Technology Association.

Chapter 9

Textiles: design and technology or art?

Chris Hughes and David Wooff

Introduction

There is a growing trend in schools to locate teaching about textiles in art and design rather than design and technology (D&T). In this chapter, we consider some of the debates around this issue and, as you might expect, we argue for textiles remaining within D&T rather than becoming art based.

There are a number of reasons why some schools are removing textiles from D&T (Hughes et al. 2011a). First, with an already packed curriculum, there is a need to rationalise the number of courses offered. This situation may be further exacerbated in England by the introduction of the English baccalaureate (Ebac), which emphasises mathematics, English, history, science and foreign languages over art, D&T and information and communication technologies (ICT). In some schools, this has led to a reduction in teaching time for D&T and for art in order to facilitate the baccalaureate subject areas. As a consequence, textiles can no longer be offered in both curriculum areas so decisions have had to be made about whether it is taught through D&T or through art.

Second, due to budgetary constraints, many subject departments have less money and have had to reduce their costs. Some have responded to this by rationalising the curriculum, for example where textiles would previously have been taught in both D&T and art it is now taught in only one area. Textiles in the art curriculum can be cheaper than textiles technology as it requires less machinery and equipment, making this a more attractive option for some.

Third, some schools seem to be experiencing more difficulties in recruiting suitably qualified textiles technology teachers than art teachers. Although data are difficult to come by as they mostly refer to 'D&T teachers' as a generic group rather than teachers within the specialist areas, there is anecdotal evidence from schools and teacher training institutions to support this. Where schools cannot recruit staff they will, of course, make the decision to no longer offer textiles technology, and pupils' only experience of textiles will be through art and design.

A further influence on the D&T curriculum has been the government decision that it should no longer be a compulsory subject at upper secondary level. In the original National Curriculum, D&T was compulsory for all pupils aged 5 to 16 and, as such, had equal status with other subjects. Over time, this has been eroded, which has led to fewer pupils studying D&T, for example between 2010 and 2011 there was a fall of almost 12 per cent in D&T examination entries for 16-year-olds (www.data.org.uk). The consequence of this is seen in the factors discussed already: smaller budgets and fewer staff.

Presented together, these are worrying trends. We believe that denying pupils access to the wealth of technological experiences textiles technology can offer will have a marked effect on the key skills, knowledge and understanding they can acquire in this prominent technological field. And although textiles provide a stimulus for aspects of an art-based curriculum, such as fashion design, fabric embellishment and fabric printing applications, this is a limited view of textiles. We take the view that textiles education should encompass the broader D&T spectrum, particularly the technological, scientific, cultural and manufacturing factors that make textile products a success in the world marketplace. In this context, a D&T approach to teaching textiles provides pupils with more opportunity to learn about all aspects of the textile design, make and distribution fields as well as the technological principles, skills and knowledge that underpin the discipline.

Textiles and the English school curriculum

Textiles teaching, in one form or other, has been a feature of the school curriculum in England and Wales for a number of years. In the early part of the twentieth century girls were taught sewing skills as part of their 'housecraft' lessons in preparation for life as a housewife or domestic servant. From the 1960s through to the late 1980s sewing formed part of the domestic science provision in most secondary schools where the broad purpose was to teach pupils, mainly girls, the rudiments of tasks such as dressmaking and sewing household furnishings.

This approach changed radically with the introduction of the first National Curriculum in 1990 (DES/WO 1989). In this curriculum, textiles technology became a component of D&T alongside systems and control, resistant materials and food technology. There was also change in the way the textile technology curriculum was taught with more emphasis being placed on technological applications (Kimbell 2004), for example how textiles linked to industrial design and manufacture, developing pupils' understanding of scientific principles related to textile applications and encouraging pupils to become aware of issues such as sustainability and design for disassembly. Textiles teaching and learning has evolved from a subject primarily linked to household applications to one that now includes a range of applications relating to the modern textile manufacturing environment.

However, even though textiles technology is now firmly embedded in the D&T curriculum, it is also a feature of art-based work in schools. In art, textiles provide a creative medium through which pupils can explore and develop artistic capabilities such as drawing and sketching, the creative use of fabric collage and fashion design.

Textiles in D&T and textiles in art share some of the same techniques and applications, for example designing with fabric materials and understanding the characteristics of fabrics for product construction. This is particularly the case where sketching and modelling are utilised in both art and D&T to create designs prior to product or artefact development. Despite these similarities, the purposes for using such techniques are markedly different. In D&T, learning outcomes are tailored towards developing pupils' product design capabilities and their understanding of the functional and aesthetic characteristics of textiles, while in art the focus is on the aesthetics and textiles provides another medium through which pupils' artistic capabilities can be developed. This fundamental difference in the focus of teaching and learning about textiles provides the basis for our argument that textiles should remain within the D&T curriculum. While we see the value of textiles as an artistic medium, textiles have an important place in modern global manufacturing and business and we think it is important that young people learn about this.

Textiles in the global economy

Textiles is a global industry in which nearly every country, culture and economy are involved in its manufacture and consumption at some level and design and manufacturing facilities for the industry are spread across almost all areas of the world. Although a significant part of the textiles sector is currently dominated by the developing countries, particularly India and China (http://ec.europa.eu/trade), all the industrialised countries, including the UK, are important producers of textiles products of all kinds. At what might be termed the lower industrial level, products such as handmade carpets are made in parts of Turkey and Iran, fabrics are woven by Inuit peoples of the Arctic region of Canada and brightly coloured cloths are made by many African tribal peoples. At the other end of the manufacturing scale, cutting-edge computer-operated machines are used to spin and weave fabrics for applications ranging from men's shirts to fibre reinforcements for aircraft parts.

The importance of the textile industry to the world economy is evidenced by the amount of global activity carried out by the sector, in recent years it has been reported that the world's consumers spend around one trillion US dollars buying clothing and other apparel goods (www.ifm.eng.cam.uk.ac). In the case of apparel, for instance, one-third of sales are in western Europe, one-third in North America and one-quarter in Asia (www.ifm.eng.cam.uk.ac). In addition, there are expanding markets in fields as varied as engineered textiles for automobile applications, woven reinforcements for fibre-based composites, geotextiles in civil engineering applications, medical textiles, carpet and other flooring products, soft furnishings and toys. Overall, clothing and textiles represent about 7 per cent of the worldwide economy.

In addition, textile designs and products can represent significant symbolic artefacts embedded in the cultural and religious histories of the people who wear or display them. In terms of the global market and increasing social diversity, the cultural and ethical significance of textiles products is an important aspect of textile design and manufacturing, which pupils can learn through textiles technology.

So, in terms of providing pupils with an up-to-date curriculum, relevant to the needs of modern economies, textile technology offers the skills and knowledge appropriate for those who will work and consume within such a rapidly changing environment. This, we argue, will be provided by textiles technology, within D&T, rather than textiles in the art curriculum.

The national importance of textiles technology

In the UK, textiles technology in its many forms, spinning, weaving, finishing, apparel manufacture and specialised textiles, has been central to the prosperity of many geographical areas. For instance, the industrial development of areas such as the northwest of England and Kidderminster in Worcestershire has been largely built around a once thriving textiles economy. A number of commentators have discussed the post-industrial implications of the ways in which the industry has had to re-invent itself in terms of niche marketing and market share (Parrish et al. 2006), increased globalisation (Tyler 2003), product and process innovations (Catling and Rothwell 2002) and changing workforces practices (Lowson et al. 2002) (Hughes and Hines 1993).

Over the years, both the products and the processes associated with textiles production have become more engineered and automated. Products such as components for Kevlar reinforced body armour are cut out using laser cutting machines, fabrics are woven on looms that have computer jacquards, and dyeing operations are often quality controlled using spectrometer methods. Moreover, in many cases the link between design and manufacture has become much more integrated leading to a quicker response in making products from the design stage. These design to manufacture innovations have led to a more globalised workplace in which textiles can be designed in one country and the design information rapidly relayed to manufacturing plants in other parts of the world. In relation to this, textiles technology in D&T is well positioned to develop pupils' creative thinking through providing opportunities to design and make products that are familiar and readily accessible to them, their understanding of the ethical aspects of design and manufacture and some of the technical and scientific aspects of textiles products.

Textiles and STEM

STEM is the integration of science, technology, engineering and mathematics and the opportunities that textiles technology provides to integrate STEM learning is a further argument for textiles teaching to be located within D&T (see Chapter 11 for a discussion on STEM). There are a number of initiatives, such as the STEM Ambassador Scheme and STEM clubs, which support the view that STEM should underpin aspects of D&T in the school curriculum and many commentators, for example Barlex (2007) and Hughes et al. (2011b) regard D&T as a main contributor to the STEM agenda.

Textiles teaching can embrace elements of materials science, colour chemistry, computer applications of design and manufacture, the exploration of mathematical shape and contour, efficient materials utilisation in planning and cutting fabric and other material structures, as well as enabling technologists to consider the economic and operation decisions that are made during one-off or batch production. Textiles technology also has the potential to relate pupils' creative capabilities to scientific and mathematical principles. Designers, for example, need to consider factors such as the properties of the fabrics used in designs, the way the product will be made, how design decisions relate to environmental issues and how economic considerations place constraints on the materials and processes used in production. Such decisions invariably involve a consideration of a range of engineering, scientific and mathematical factors.

Relevant textiles technology teaching and learning can ensure a strong STEM focus in D&T, especially in relating aspects of textile design activity to the technological, scientific and mathematical principles that affect the way products are made and perform in service.

Textiles in design and technology

The development of D&T in the UK schools' curriculum has, in the main, been in response to changing economic and technological needs (Penfold 1997). Since the late 1960s and early 1970s D&T has been in the vanguard of a number of progressive and technological pedagogic developments, for example pupil-centred learning, the problem-solving approach, creative thinking and its application and matching pupils' capabilities to technological thinking.

Textiles technology teaching and learning has been part of this development and has been enhanced by the introduction of innovative tools and equipment. Over the last 20 years or so textiles technology classrooms and studios have moved from largely hand tool and manually operated machines to ones which include facilities for computer-aided design activities and a range of up-to-date computer-operated machinery such as computer-aided sewing and embroidery machines and laser cutting machines. These developments have been important in linking school-based work with applications in industry and business. For example, pupils can act as designers in much the same way textile designers operate in practice, they can make prototypes in the same way as many products are made at the industrial level, make mock-ups using low-level tools and equipment and evaluate and modify designs ready for production. In this way, pupils acting as designers can take into account a range of factors such as drape and properties of materials, how the fabric might behave when in use, how it can be effectively manufactured and how economics can come into the planning of the work.

In contrast to the approach taken in art, textiles technology teaching is focused on helping pupils understand the functional properties of traditional and modern textiles and their wide range of uses in the modern world. Pupils are taught:

- classification of materials by their fibre sources: natural, synthetic and regenerated
- fabric construction techniques, from fibres to fabrics
- the cultural contexts in which textiles are developed and used
- colouring and decorative techniques
- the aesthetic and functional properties of materials, e.g. water resistance, drape, comfort, absorbency and flexibility, and the connection between the properties of fibres, their performance and how they are used, e.g. cotton is cool to wear, which makes it ideal for knitted T-shirts
- how the properties and working characteristics of textile materials may relate to their composition, construction and finishing and how synthetic fibres can be made to emulate natural fibres, e.g. microfibres
- sustainability in relation to textiles
- some of the industrial uses and applications of textiles.

In textiles technology, pupils develop technology skills, knowledge and understanding, they develop their creative skills and develop an understanding of the close relationship between the design aspect and the technical aspects of work in the profession.

Practical work can focus on making and designing for one's self or for clients. Pupils can learn pattern techniques, such as jigs and templates, make mock-ups of their designs, construction techniques and decorative techniques. They can learn about manufacturing processes, such as one-off and batch production, and the use of ICT in design and manufacturing.

Textiles products are also an excellent way to help pupils explore values and ethical issues in design and manufacturing of products. Many garments and other textiles products are manufactured in countries where wages are low and there are poor conditions of work, using techniques that would breach health and safety standards in developed countries. There are also issues surrounding the use of raw materials, such as pesticides in the growing of cotton, the use of harmful dye techniques to colour textiles and the use of unregulated raw materials in some countries to manufacture products. Examining the way products are manufactured throughout the global supply chain can lead to raising pupils' awareness of ethical considerations in the design and manufacturing process and how their decisions as consumers can impact on the lives of others in other parts of the world.

Textiles in art

We have already made reference to the similarities between textiles in art and textiles in D&T in the school curriculum. Both subject areas emphasise pupil learning through the design and application of fabrics, encourage creative thinking and discovery during learning activities, develop pupils' competence in making textile artefacts and encourage pupils to explore available techniques and processes using ICT methods such as laser cutting. However, the two subjects each take a fundamentally different approach. These differences lie not so much in the end product, whether it is an item of clothing or piece of household furnishing, but in the approach to learning and the intended outcomes. In these respects, art and D&T are different in relation to their pedagogic purposes, learning outcomes and assessment criteria.

At the heart of the art-based approach, textiles teaching is focused on enabling pupils to explore and develop a range of techniques that relate to the manipulation of fabrics including stitch-based practices, fashion illustration, embroidery, woven art work and three-dimensional art constructions. To achieve these outcomes pupils often use techniques such as free machine embroidery, hand embellishment of fabrics, they print designs on fabric materials and make use of artistic methods of fabric adornment and decoration. Typical art-based activities in the lower secondary school include examining textiles in terms of their colour, weave and weight and using such information to create artefacts in moulded, three-dimensional or woven forms. Dyeing, printing and painting on fabrics features prominently in many schools' schemes of work where pupils learn through activities such as making wall hangings, staining and bleaching fabrics to change their appearance. They also become familiar with various techniques such as batik and screen printing to create patterns and images. Thus, art-based textiles may be seen to intensify pupils' perceptions of the world around them by manipulating fabric materials in an artistically creative way.

The outcomes of textiles-based art work is for pupils to be able to work confidently and creatively with textiles in order to express themselves. And the assessment criteria for this focus on pupils demonstrating their knowledge and understanding of art techniques and art forms and the ability to create, record and make personal and meaningful art pieces.

This is a valuable contribution to pupils' education, the development of creativity and personal understanding is a worthwhile endeavour. However, it should be seen as complementary to the textiles work undertaken in D&T, not a substitution.

Textiles in school: art or technology?

As outlined earlier, textiles teaching and learning, in one form or another, has been a feature of the UK's secondary school curriculum for a number of years. Generations of pupils have developed their understanding of working with textiles through the design and manufacture of artefacts such as apparel, soft furnishings and art features. Textiles have also held a central position in the UK economy. Many communities in the UK have been shaped by the work patterns and behaviours of those who work in the textiles industry, and textiles design and manufacture can be seen to have made significant contributions to the prosperity of the nation. As discussed, textiles industries in their broadest sense, including all aspects of design, manufacture and distribution, account for a large proportion of global economic activity and most countries in the world are engaged in textile manufacture to a lesser or greater degree. In many cases, the industry is characterised by highly automated processes to facilitate both the design and manufacturing elements of the industry, helping make textile design and technology a key driver in the UK and global economy.

All cultures make use of textiles products of some kind. In the more advanced nations, textiles use and consumption accounts for a considerable proportion of economic output. Products include flexible forms of textiles, for example menswear and women's clothing as well as engineered products such as fibre reinforced aircraft parts and geotextile products that help reduce soil erosion. In this way, we see textiles as being fundamental to many countries' standard of living, economic well-being and social standing within the global economy. It is textile technology's intimate relationship with aspects of our lives and working practices that, we believe, make it a central component of an up-to-date and relevant schools' curriculum. However, what does the future hold?

As we said at the start, there is a shift in the UK schools' setting towards offering an arts-based textiles curriculum in preference to a design and technology one. However, the two textiles routes have markedly different aims and outcomes. For example, arts-based textiles tend to focus on aspects of shape, colour, texture, pattern appreciation, harmony and an appreciation of form and function. There is minimal focus on the technical aspects of the subject, which may include, for example, an understanding of batik, appliqué and fabric printing.

Textile technology, by way of contrast, generally has a product design focus that includes technologically oriented applications. These include appreciating the relationship between design, materials, manufacturing and marketing; understanding the technological use of fibres; methods of textile manufacture; designing with a human purpose; understanding environmental concerns; the use of CAD/CAM in textiles production; and elements of pattern drafting.

Both artists and design and technologists create visual artefacts and products using a shared knowledge base, but their reasons for doing so are usually entirely different. Typically, from the artist's point of view, the process of creating a piece of art stems from an opinion, view or feeling that the artist holds within. Thus, it might be argued that the artist creates their work as a vehicle for sharing their feelings with others and the approach or purpose of arts-based textiles work is to encourage 'open' experimentation using textiles as a medium for expression. Design and technologists, in contrast, begin their process of creating based on what is described in a structured design brief. This helps frame the parameters for the design, manufacture and distribution factors that influence the design activity. We would also argue that this functional, problem-solving approach to design also develops pupils' creativity in a much more challenging way. The design has to be creative but also functional and appropriate.

Our support for textiles teaching and learning remaining firmly within D&T is borne out by comments made in an OfSTED (www.ofsted.gov.uk) report. Here it was noted that there were insufficient opportunities in the curriculum for pupils to develop knowledge in areas such as advanced design and manufacturing. The report suggested that this was a key weakness at a time of rapid technological advance and that pupils have minimal teaching and learning opportunities to learn about innovative technological techniques that combine a scientific understanding of design when making practical products and systems. Such technological understanding can only really be achieved through pupils engaging with a textiles course of study that incorporates a strong technological focus; a focus that needs to develop pupils' capabilities through a design and technology programme of study.

Questions

1   What do you perceive to be the differences between textiles taught through art and textiles taught through D&T?
2   If schools do have to make a choice, which subject area do you think is most appropriate for teaching textiles? Why?
3   What do you think would be the consequences if textiles technology were removed from the D&T curriculum?

References

Barlex, D. (2007) *'Capitalising on the utility embedded in design and technology activity: an exploration of cross curricular links'*. Paper presented at DATA Education and Research Conference, Wolverhampton.

Catling, H. and Rothwell, R. (2002) 'Automation in textile machinery', *Research Policy* 6(2): 164–176.

DES/WO (1989) *Proposals of the Secretary of State for Education and Science and the Secretary of State for Wales. Design and technology for ages 5 to 16*, London: HMSO.

Hughes, C., Bell, D. and Wooff, D. (2011a) 'Reducing the practice gap between the design and technology curriculum and the needs of the textile/manufacturing industry' in Perspectives of Learning in Design and Technology Education, PATT 25/Cript 8 Conference, 1–5 July 2011.

Hughes, C., Bell, D. and Wooff, D. (2011b) 'Underpinning the STEM agenda through technological textiles: an exploration of design and technology teachers' attitudes', *Design and Technology Education: An International Journal* 16(1): 53–61.

Hughes, C. and Hines, T. (1993) 'Technology and innovation for the 21st century in fashion design' in International Conference in Fashion Design, University of Industrial Arts, Helsinki.

Kimbell, R. (2004) 'Design and technology' in White, J. (ed.) *Rethinking the school curriculum: values, aims and purposes*, London: RoutledgeFalmer.

Lowson, R., Hunter, A. and King, R.E. (2002) *The textile/clothing pipeline and quick response management*, Chichester: John Wiley & Sons.

Parrish, E.D., Cassill. N.L. and Oxenham, W. (2006) 'Niche market strategy in the textile and apparel industry', *Journal of Fashion Marketing and Management* 7(3): 231–234.

Penfold, J. (1997) 'From handicraft to craft design and technology', *Studies in Design Education, Craft and Technology* 20(1): 34–47.

Tyler, D. (2003) 'Will the real clothing industry please stand up!', *Journal of Fashion Marketing and Management* 7(3): 231–234.

Chapter 10

Using technology in design and technology

Alison Hardy and Sarah Davies

Introduction

Technology is ubiquitous in the modern western world and throughout the history of design and technology (D&T) in the school curriculum technology has been synonymous with its content and teaching. In discussing 'technology' in the D&T curriculum, we are focusing to the use of computer-based equipment for the purposes of designing and making. There are several debates around the use of technology in D&T classrooms; in this chapter, we focus on three issues: the purposes of technology in D&T, how teachers use technology to develop pupils' learning and the availability and management of technology resources.

In discussing the purpose of technology in the D&T curriculum, we consider how technology can support the development of pupils' industrial knowledge and their skills of creativity and product manufacture. In the second debate, we consider the integration of technology into D&T classrooms through effective teaching strategies and how a teacher's ability to be effective will be influenced by both their level of subject confidence and their personal view of the use of technology in the classroom. The final debate discusses some of the pragmatic issues around the purchase and location of equipment, showing how resource management can impact on learning.

To support you in your journey through this chapter we have established the scope of technology to be considered (see Table 10.1). From the vast range of technologies we could have considered, we have chosen to focus on three uses of technology within D&T classrooms: industrial design and manufacturing technologies, mobile technologies and online technologies.

Background

The first National Curriculum in the UK, in 1990, brought together a number of previously discrete subjects, including information and communication technology (ICT), into a new subject called technology. At that time, ICT within technology focused on word processing, databases and spreadsheets with some computer-aided design (CAD), programming and control. As the use of

Table 10.1 Scope of technology in D&T considered in this chapter

Name	Definition	Examples	Purpose/ uses
Industrial design and manufacturing technologies	Computer-aided design (CAD) Computer-aided manufacture (CAM)	CAD: Techsoft 2D Design, Microsoft Publisher, SpeedStep and ProEngineer CAM: computerised sewing machine, tunnel ovens, food processors, knife cutting machines, milling machines and laser cutters	CAD: used for drawing two- and three-dimensional images CAM: used for manufacturing either parts or whole products and artefacts
Mobile technologies	Handheld technologies	Tablet computers, mobile phones and digital cameras	Voice recordings and photographing 'dirty models'(Bramston 2008) to capture key decision moments in the design process. 'Apps' on smart phones used for materials research and revision. Tablet computers have 'apps' for drawing and design
Online technologies	Websites, Web 2.0 technologies and content management systems	Search engines, such as Google, websites found through searches, PBworks, Moodle, YouTube and Slideshare – and many, many more	Websites used for research, collecting information. PBWorks and Virutal Learning Environment (VLE) discussion boards used by groups to collaborate on projects, developing their own understanding and knowledge

computer-based equipment grew in D&T, partly as a response to the use of computers in industry and society, there was an increasing need for investment in both resources and training in these new technologies.

In 1999 the Design and Technology Association encouraged the use of CAD/CAM (computer-aided design and computer-aided manufacture) in schools through promoting design software (e.g. ProDesktop and SpeedStep) and CNC (computer numerical control) machines, together with providing training for teachers. This, along with a greater emphasis in the curriculum on pupils' learning about industrial practices, created a significant growth in the use of computers and a need for schools to invest in technology.

In later revisions of the National Curriculum, ICT was removed from D&T, which gave teachers greater freedom to select technology resources and content that were appropriate for D&T rather than for general ICT learning. In 2007 this became more of an imperative, with the National Curriculum stating that in D&T, pupils 'learn to use current technologies and consider the impact of future technological developments' (QCA 2007: 51). It was a requirement of the programme of study that they learn to 'evaluate which hand and machine tools, equipment and computer-aided design/manufacture (CAD/CAM) facilities are the most appropriate to use' (ibid: 54), and specifically stated that they should use ICT when designing. It suggested that curriculum opportunities to use ICT could include:

> image capture with scanners and digital cameras; image generation through computer-aided design (CAD); data acquisition through CD-ROMs and internet-based resources; data capture through sensors; data handling through the use of databases and spreadsheets; controlling through the use of control programme software; and product realisation through the use of computer-aided manufacture. (CAM) (ibid: 57)

However, the UK government is increasingly concerned about the role of ICT in schools and is proposing to introduce 'computer science' into schools to replace ICT as it is currently taught. This could have an impact on D&T as there will be an increased emphasis on programming, which already forms part of the D&T electronics curriculum in many schools. Any review of ICT/computing teaching could lead to beneficial developments in D&T teaching, if its role is recognised, or negative developments if programming is placed elsewhere on the curriculum and not seen as a relevant part of D&T teaching. The future is uncertain and teachers need to be arguing for the relevance of D&T to the teaching of computer knowledge and skills.

If D&T is going to remain an important part of teaching pupils about technologies then it needs to address the issues of the appropriateness of the technologies used, the learning that takes place when pupils use technology in D&T and how the resources are managed.

Purpose of technology in D&T

Technology is often assumed to be an integral part of the D&T curriculum, partly due to the history of the subject as outlined already (and in more detail in Chapter 1). In this debate, we present some justifications for its inclusion. However, in justifying the use of technology in D&T, we also need to consider the rationale for D&T on the school curriculum as each rationale provides a different justification. There are different views on the value of D&T in pupils' education, here we focus on three, these are that pupils should learn D&T because it:

1   teaches about the real world of design and manufacturing-related industries
2   develops pupils' creative design skills
3   develops pupils' practical skills.

These three views are not independent of one another but each influences differently the way technology is taught and used in D&T.

The first view, which many consider to be the primary purpose of D&T as a school subject, emphasises the importance of informing pupils about, and preparing them to work in, design- and manufacturing-related industries. This view supports the justification of CAD/CAM in D&T as it brings the 'real world' into the classroom through the simulation of industrial practices. Steeg (2008: 1) suggests, in addition, that the D&T curriculum is founded on 'developing in pupils designing and making skills and knowledge that are derived from industrial design'. Teachers holding this view are likely to teach design process models from industry and encourage pupils to learn industry-based knowledge and skills. They will support the use of technology equipment such as CNC machines, computerised sewing machines and tunnel ovens that allow pupils to replicate batch manufacture and create multiples of the same product.

Examination specifications also, in part, subscribe to this view; they require pupils to demonstrate an understanding of the use and application of computers in design, manufacturing and engineering. This requires teachers to include teaching about these applications of technology from an industrial perspective.

The second view of the value of D&T education, that it develops pupils' creative design skills, is reflected in the 2007 version of the English National Curriculum importance statement for D&T, which states that it allows pupils to 'apply their creative thinking and learn to innovate' (QCA 2007: 51). There are several ways in which technology can be used to foster and encourage pupils' creativity, for example accessing online information, such as blogs, e-magazines and wikis, to better understand the context for design work or using online images as a design stimulus. These images can provide different and unusual stimuli for pupils' designs, which can lead to more creative solutions (Nicholl and McLellan 2008).

Technology can also help pupils to explore and develop their design ideas (Hodgson and Fraser 2005), for example through using CAD to alter designs, produce variations and modifications and manipulate shapes and materials. CAD can also be used effectively by pupils who have difficulty in drawing and who struggle with presenting their design ideas through freehand sketching. Pupils can 'cut and paste' 2D drawings from the internet and manipulate them into a new shape, so using CAD software might enable a pupil to produce something, and be more creative, than they would have been using only pencil and paper. But is this an appropriate use of technology; should CAD be used in place of freehand sketching?

Sketching with pen or pencil requires practice and, with reduced curriculum time available for D&T, some teachers may be persuaded to replace teaching sketching skills with CAD. But CAD has its limitations, for example it cannot generate quick 'back of the envelope' sketches and even with programs such as CREO Sketch, children still need time to be able to use the software with confidence. Banks and Owen-Jackson (2007) suggest that the time needed to become competent in using CAD is shortened as the software develops but, as yet, there is little evidence-based research to support this suggestion. Research from Coyne, Park and Wiszniewski (2002: 271) suggests that 'inexperience *seems* to limit design possibilities', implying that pupils' ability to be innovative when designing using CAD might be limited by their lack of competence in CAD, just as it would be if they lacked competence in drawing skills. This view might be challenged, however, by teachers' classroom experiences which often show that after demonstrating the basic principles of a CAD program, some pupils will experiment and take risks with the software to realise their design ideas.

Jonson (2005) suggests it is the verbalisation of ideas that is common to both drawing with a pencil or CAD, so if teachers encourage pupils to talk about their designs as they develop them this combines two powerful tools – discussion and drawing. Mobile technologies can be effectively used to support and capture the development of ideas that comes about when pupils combine talking with drawing and modelling their ideas either on paper or a computer. The images and files collected using mobile technology can be compiled into an e-portfolio, which pupils present to show their design thinking.

Other technology that can help with design development and recording includes digital pens, which can record handwriting, graphics and voice recordings; digital recorders for voice recordings; mobile phones and cameras for taking pictures, videos and voice recordings. All these, we would argue, complement the pupils' learning and development of designing; they do not replace other skills but extend them. These technologies also give pupils the opportunity to record their thinking as it is happening and making it available for use later, for example reflection for an e-portfolio.

E-portfolios can be produced in real time using mobile technologies such as tablets and mobile phones. This enables pupils to capture their design development, going back and forth over ideas in real time, creating real development in their design ideas in contrast to paper-based portfolios where 'development' often means tidying up rough sketches to make them look nice. Using mobile technologies to capture design decisions and development as they occur has been explored in the E-scape project (E-Solutions for Creative Assessment in Portfolio Environments; http://www.gold.ac.uk/media/e-scape_phase3_report.pdf).

The third view of the value of D&T is that its purpose is to develop pupils' practical skills. In this view, technology, particularly CAD/CAM and remote making, are tools that can improve the quality of pupils' practical work, enabling them to 'achieve a professional quality' (OfSTED 2011: 33). Schools often display pupils' work in order to showcase the high-quality hand-crafted work that has been produced. As products can now be designed by pupils in school then manufactured remotely using machines controlled by computers, some teachers feel that craft skills are being lost. How appropriate are hand-craft skills, how appropriate is the use of technology for making, do we need to redefine 'practical skills' in twenty-first-century D&T?

There is still a common understanding that D&T is about 'making things' and there is pressure on teachers to support pupils in making products that are 'highly professional and well designed (OfSTED 2011: 31). But what does 'highly professional' mean in this context? Despite a current (2012) revival of interest in hand-made products and DIY hobbies, pupils are very aware of the aesthetics of products and want to make products that look 'good'. This means that they need to be well made with a good-quality finish, comparable with products available in the shops. Technology provides the ability for pupils to do this, for example laser cutters make it easier to cut sheet materials with accuracy, enabling pupils to make well-finished products of which they are proud. But what does the pupil learn from this experience? Are we developing pupils' ICT skills, their D&T skills or something else?

This section has looked at three different views of the purpose of D&T education and how the view a teacher holds will determine in part the way in which technology is used in the classroom, or even if it is used at all. When teachers do use technology in D&T, the pedagogical approach they choose affects how the technology is used and what the pupils learn.

Using technology to develop learning

Teachers develop teaching strategies that they consider to be effective in making clear to pupils what they are doing and why they are doing it. In the case of using technology, we need to consider how teachers' ability to be effective will be influenced by their level of confidence as well as their personal philosophy of education.

Teachers are often defined by their subject knowledge and by this we mean subject content knowledge, for example what they know about materials or ingredients, product manufacture and design processes. This subject content knowledge may derive from university studies or professional, industrial experiences and it has been shown to be linked to teachers' level of confidence in the classroom (Banks, Leach and Moon 1999). While most teachers qualifying to teach today are likely to be 'digital natives' (Prensky 2001) and therefore comfortable with CAD/CAM, handheld and web-based technologies, this cannot be assumed for all teachers and it does not acknowledge the rapid rate of change in technology, which leaves even the most up-to-date teacher with incomplete knowledge.

The nature of D&T means that teachers are (usually) continually acquiring new subject knowledge, whether through formal training, experience or learning from colleagues. This knowledge, though, is often learnt without consideration of any associated pedagogy, that is how the new knowledge will be taught or how the pupils will learn. Developing new pedagogies, or adapting existing ones, can be time consuming and can create tension for a teacher between spending time building confidence in the use of the new technology or spending time developing teaching strategies to support effective integration of it in the classroom.

Research shows that, as a result of low skill level and low confidence, teachers often start out teaching didactically when introducing a new technology or software program (Sandholtz et al. 1997). For example, when introducing a new CAD program, the teacher may adopt the stance of instructor and an over-reliance on the free tutorials provided with the new software. With such didactic teaching, pupils may learn one way of using the software but would be limited in their ability to develop its potential for creative exploration. A more skilled or confident teacher could develop a pedagogical approach focusing on the potential of the software for allowing new ways of working, such as giving pupils opportunities to work out for themselves how to draw and test their ideas on screen, with little direct teacher input. One teacher (Winn 2012), confident in teaching 3D designing software and tired of the standard step-by-step approach, created a game approach to learning in which the pupils had to learn elements of the software in order to solve problems and progress through the levels of a game. The pupils were more enthusiastic about learning and developed a deeper understanding of the software than those taught by the usual methods. Using such an approach the pupils become confident in using the software and so more likely to use it through their own initiative later.

Developing knowledge and skills in new technologies is an ongoing process for teachers and their pupils and when they see it developing together this can support both teachers and pupils in getting the best out of the technology. As with other aspects of D&T, if the teacher can model for the pupils what it is to learn about the new technology as they do it with them, it can be beneficial for pupils' learning. Table 10.2 depicts a five-stage model of teacher and pupil progress in pedagogy and learning towards a socially interactive and reflective approach. This was developed from the work of Sandholtz et al. (1997) and shows the teaching and learning strategies that a teacher may draw on in a framework of five phases: entry; adoption; adaption; appropriation; and invention.

The five phases described in Table 10.2 create a framework that can be used by teachers to reflect on their learning, and that of their pupils, when developing knowledge of the use of new technologies in the classroom. The table shows the key characteristics for pedagogy and resource management from the pupil and teacher perspective. Links are made between phases to support teachers in reflecting on where they, and their pupils, are in the process of developing subject knowledge of new technologies.

A teacher's personal philosophy of education also has an influence on pupils' technology learning, so teachers need to be aware of their own philosophy and how it impacts on the choices they make within the classroom. We have identified three characteristics, shown in Figure 10.1, that contribute to the teacher's personal philosophy.

Table 10.2 Stages of teaching and learning with technology in D&T

Phase	Description	Teaching/ learning	Resource/ classroom management	Transition to next phase
Entry	This is when the technology is new to the teacher and/or pupil.	Teachers are excited and apprehensive, with little experience or no experience of the technology. The teacher is unable to anticipate problems and becomes frustrated. Both the teacher and pupil make mistakes and this can be a significant deterrent to further use of the technology.	Discipline and resource management difficulties occur as they are unanticipated – similar to those experienced by newly qualified teachers when they first start teaching.	After a period, however, the teacher gains in confidence and is able to move on in her use of ICT.
Adoption	The technology is beginning to be integrated into current projects and schemes of work by the teacher and/or pupil.	The teacher spends time evaluating the technology with respect to established projects and teaching strategies. The teacher's lack of experience in the technology hinders pupil progress. Pupils are pressured to succeed in basic skills rather than exploring the technology and its potential.	Problems with the management of the classroom and resources are anticipated and planned for.	A teacher's pedagogical approach begins to change to allow pupils to experiment more.
Adaption	The technology is seen as a support for the curriculum rather than an infringement and is fully integrated into traditional classroom practice.	Traditional pedagogy is relied on such as lecturing, seated work and individual quiet task. As the pupils become more confident and assertive in their learning with the technology they become curious and take on new challenges beyond the remit of the lesson. Use of it is more purposeful.	Classroom management style remains the same, pupils are working individually with the technology.	Pupil pride in the work develops and towards the end of the adaption stage the classroom environment changes with a shift from a teacher-led approach to a more pupil-centred environment.

Phase	Description	Teaching learning	Resource/classroom management	Transition to next phase
Appropriation	This is seen as a milestone as it is about the personal attitude of the teacher towards the technology. The pedagogy and technology begin to be considered as a whole.	A teacher's personal philosophy is changed and the integration of the technology into teaching and learning is instinctive. Pupils 'use it effortlessly as a tool to accomplish real work' (Sandholtz et al., 1997: 42).	Departments consider the location and accessibility of the resources to support independent selection by pupils of the technologies as opposed to being directed by the teacher.	Pupil expectations and teacher's awareness of the potential for new pupil competencies drives a need for further change in pedagogy.
Invention	The key to reaching the stage of invention is teacher reflection, questioning old patterns and speculating on the change seen in their pupils when using technology.	Teachers are experimenting with new pedagogy that supports deep learning. Pupils are collaborative and the tasks are interactive. The learning is more pupil centred, with shared responsibility for teaching and learning with technology. The teacher views knowledge as something constructed by the pupils for themselves.	Classroom management is intertwined with teaching and learning strategies that focus on collaboration and pupil-centered learning.	

(based on Sandholtz et al. 1997: 37–47 and 56–75)

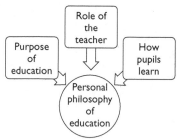

Figure 10.1 Model of a teacher's personal philosophy of education

Loveless et al. (2001) reflect that the teacher should think about how ICT (technology) might support or alter their approaches to teaching strategies, in the context of how pupils learn. For example, if teachers hold a constructivist view of learning, which is that pupils learn by constructing their own understanding of the world, then they would develop pedagogies which allow for pupil autonomy. In their article, Loveless et al. assume that all teachers want their pupils to become autonomous learners and that, in order to do this, the teacher's role should be one of mentor or facilitator rather than instructor. This means that the teacher needs to consider both the role they will play in the lesson, for example director of activities, facilitator or lecturer, and the learning they want to promote (Scrimshaw 1997). The teacher's role will vary depending on the learning situation, sometimes the teacher is unobtrusive, at other times a hands-on director of activities. So teachers need to be prepared to develop their practice and try different ways of working, using the five-stage model in Table 10.2 to reflect on their progress.

What pupils learn from using technology in D&T depends, to a large extent, on the teaching strategies deployed by the teacher. They can learn by rote procedures for using equipment or software or they can learn how to 'play' with technology to find its potential and value.

Resource management

No teacher or school can consider the use of any technology without giving consideration to its cost. Where resources are limited, consideration also needs to be given to their availability and accessibility and the associated classroom management issues.

Design and technology is an expensive subject in schools because of the cost of buying and maintaining equipment. The need to update this equipment, especially relevant when considering ICT, adds to this. However, these costs cannot be avoided as there are curriculum requirements for the use of industrial design and manufacturing technologies and examination specifications make reference to the use of CAD/CAM, either requiring pupils to understand industrial applications or requiring coursework in which pupils must use CAD/CAM. This places an expectation on schools to provide appropriate equipment.

Phase	Description	Teaching/ learning	Resource/ classroom management	Transition to next phase
Appropriation	This is seen as a milestone as it is about the personal attitude of the teacher towards the technology. The pedagogy and technology begin to be considered as a whole.	A teacher's personal philosophy is changed and the integration of the technology into teaching and learning is instinctive. Pupils 'use it effortlessly as a tool to accomplish real work' (Sandholtz et al., 1997: 42).	Departments consider the location and accessibility of the resources to support independent selection by pupils of the technologies as opposed to being directed by the teacher.	Pupil expectations and teacher's awareness of the potential for new pupil competencies drives a need for further change in pedagogy.
Invention	The key to reaching the stage of invention is teacher reflection, questioning old patterns and speculating on the change seen in their pupils when using technology.	Teachers are experimenting with new pedagogy that supports deep learning. Pupils are collaborative and the tasks are interactive. The learning is more pupil centred, with shared responsibility for teaching and learning with technology. The teacher views knowledge as something constructed by the pupils for themselves.	Classroom management is intertwined with teaching and learning strategies that focus on collaboration and pupil-centered learning.	

(based on Sandholtz et al. 1997: 37–47 and 56–75)

134 Debates within design and technology

Figure 10.1 Model of a teacher's personal philosophy of education

Loveless et al. (2001) reflect that the teacher should think about how ICT (technology) might support or alter their approaches to teaching strategies, in the context of how pupils learn. For example, if teachers hold a constructivist view of learning, which is that pupils learn by constructing their own understanding of the world, then they would develop pedagogies which allow for pupil autonomy. In their article, Loveless et al. assume that all teachers want their pupils to become autonomous learners and that, in order to do this, the teacher's role should be one of mentor or facilitator rather than instructor. This means that the teacher needs to consider both the role they will play in the lesson, for example director of activities, facilitator or lecturer, and the learning they want to promote (Scrimshaw 1997). The teacher's role will vary depending on the learning situation, sometimes the teacher is unobtrusive, at other times a hands-on director of activities. So teachers need to be prepared to develop their practice and try different ways of working, using the five-stage model in Table 10.2 to reflect on their progress.

What pupils learn from using technology in D&T depends, to a large extent, on the teaching strategies deployed by the teacher. They can learn by rote procedures for using equipment or software or they can learn how to 'play' with technology to find its potential and value.

Resource management

No teacher or school can consider the use of any technology without giving consideration to its cost. Where resources are limited, consideration also needs to be given to their availability and accessibility and the associated classroom management issues.

Design and technology is an expensive subject in schools because of the cost of buying and maintaining equipment. The need to update this equipment, especially relevant when considering ICT, adds to this. However, these costs cannot be avoided as there are curriculum requirements for the use of industrial design and manufacturing technologies and examination specifications make reference to the use of CAD/CAM, either requiring pupils to understand industrial applications or requiring coursework in which pupils must use CAD/CAM. This places an expectation on schools to provide appropriate equipment.

An indication of some of the costs of industrial manufacturing machines is given in Table 10.3, but it is worth noting that some of the lower cost machines only simulate industrial practices while the higher cost ones may be found in manufacturing businesses.

As with the purchase of any teaching resource, especially ICT, teachers and schools need to consider the educational value that it would bring, how it would be used and how their teaching pedagogies may need to adapt to make best use of the resource.

The underlying educational philosophy of the purpose of D&T on the school curriculum may also influence purchasing decisions. For example, if the D&T department believes that its main focus is teaching pupils about the real world of design and manufacture, it would consider purchasing machines that are used in industry, such as laser cutters and milling machines that would help pupils create professionally made products. It is sometimes the case, however, that the available budget determines what can be purchased and low-cost alternatives are still able to provide pupils with experience of CAM equipment.

Once equipment has been purchased, decisions have to be made about location, accessibility and availability, as where equipment is located and how accessible it is to pupils can have a significant effect on their learning. A report from the Design and Technology Association (DATA 2007: 72) reported that 'the lack of adequate resources and updated accommodation ... influences teaching and learning' and OfSTED (2011: 31) found, from inspection of 89 schools, that 'Where ICT, and particularly CAD and CAM, were readily available to support designing and making, they made a good contribution to students' learning.'

Table 10.3 Cost of ICT equipment

Equipment	Cost	Comment
2D cutter	Starts at £300 up to £12,000	Lower cost machines will cut card, paper and vinyl. Higher cost machines will cut plastic or larger sheets of vinyl.
Laser cutter	£10,000 to £30,000	Can be used in resistant materials and textiles. Higher cost machines are faster and more precise.
Basic CNC milling machines and routers	£3,000 to £15,000	Lower cost machines limit the size and type of material that can be cut. Higher cost machines are more versatile.
Computerised sewing and embroidery machines	£500 to £4,000	Lower cost machines require manual changing of thread when there is a colour change. Higher cost machines can use up to 10 different colours.
Cake printer	£3,000+	Uses edible inks to print 2D CAD designs onto a cake or other food item.

One issue for consideration is the accessibility of resources. Where equipment is expensive schools often restrict to buying one item; if this is then located in a storeroom and operated only by the teacher or technician this limits pupils' learning. Alternatively, it may be located in the workshop or classroom with the most space, which means that it is more likely to be used by the teacher and pupils who occupy that room than by others, again limiting the experiences and learning of some pupils.

Computers are usually located either together in one room or spread among classrooms. Where they are located in one room, this often leads to whole-class teaching, which may not always be appropriate or necessary, or individual pupils working away from the rest of the group, again not always appropriate. Where they are spread among different rooms, there are fewer computers available but pupils have access to them as and when they are needed. Where schools can afford both, a computer room and computers in classrooms, this provides the optimum solution.

Where computers are available they allow for the use of online technologies and web-based resources. Many web technologies are free to education, for example wikis, which can be used to collaborate on group projects and websites for sharing resources and blogs.

The use of ICT in D&T is also changing with the increase in the availability and accessibility of pupil-owned mobile technologies such as smartphones and tablets. Although some schools ban these, on the grounds of safety, security, distraction and fear of the digital divide, i.e. inequality in the classroom when not every pupil has a smartphone or mobile phone to bring into school, How (2011: 14) states that the 'impact of allowing pupils to use their smartphones and tablets in lessons has been shown to be high'. In D&T, mobile technologies have tremendous potential to aid pupils' creativity and innovation through allowing them to generate, develop and record their design decisions in real time.

The organisation and management of technology resources is becoming an ever more important consideration. Thought has to be given to what can be purchased, where it is to be sited, who will have access to it, how it will be used or integrated into the curriculum and what value it will bring to pupils' learning. Decisions about using technology in D&T are not easy.

Conclusion

This chapter has considered the use of technology in D&T and has identified that using modern technologies is not straightforward. Consideration has to be given to the rationale for teaching the subject, the purposes for using technologies rather than other methods of teaching and organisation and management issues.

The chapter began by considering different views of the purpose of D&T education and how the view that a teacher holds influences the use of technology in the classroom. It then discussed technology-related pedagogy and the potential technology has for promoting independent learning. Finally, consideration was given to the affordability, accessibility and availability of technology resources.

At the beginning we stated that technology is synonymous with D&T and teachers have positively responded to the constant evolution of new technologies; however, this has taken place sometimes without careful consideration of the purpose and, therefore, spread of the technology. Technology has the potential to limit pupils' D&T experience, if it is restricted or taught didactically, and it has the potential to make D&T an exciting, challenging, contemporary subject. We recognise that there will be new technologies this year, next year and into the future that we cannot even imagine, but we believe that we have provided here a framework for teachers to debate the value of known and as yet unknown technologies that can and even should be integrated into D&T education.

Questions

1 What is your personal justification for the use of different technologies within your classroom?
2 How do your teaching and learning strategies when using technologies support pupil autonomy?
3 How does the location and access to technology support pupil-centred learning in your department?

References

Banks, F., Leach, J. and Moon, B. (1999) 'New understandings of teacher's pedagogic knowledge' in Leach, J. and Moon, B. (eds) *Learners and pedagogy*, London: Paul Chapman.
Banks, F.R.J. and Owen-Jackson, G. (2007) 'The role of making in design and technology' in Barlex, D. (ed.) *Design and technology for the next generation*, Whitchurch: Cliffe & Company.
Bramston, D. (2008) *Idea searching*, Worthing: AVA Academia.
Coyne, R., Park, H. and Wiszniewski, D. (2002) 'Design devices: digital drawing and the pursuit of difference', *Design Studies* 23(3): 26–286.
Design and Technology Association (DATA) (2007) *Survey of provision for design and technology in schools in England and Wales (1996–2006)* (report), Wellesbourne: Design and Technology Association.
Hodgson, T. and Fraser, A.S. (2005) 'The impact of computer-aided design and manufacture (CAD/CAM) on school-based design work' in Norman, E.W.L., Spendlove, D. and Grover, P (eds) 'Inspire and Educate', The Design and Technology Association International Research Conference, Wellesbourne: Design and Technology Association.
How, B. (2011) *The importance of technology: the technological revolution in schools*, London: Schools Network.
Jonson, B. (2005) 'Design ideation: the conceptual sketch in the digital age', *Design Studies*, 26(6): 613–624.
Loveless, A., DeVoogd, G.L. and Bohlin, R.M. (2001) 'Something old, something new... is pedagogy affected by ICT?' in Loveless, A. and Ellis, V. (eds) *ICT, pedagogy and the curriculum*, London, RoutledgeFalmer.

Nicholl, B. and McLellan, R. (2008) 'Oh yeah, yeah you get a lot of love hearts. The Year 9s are notorious for love hearts. Everything is love hearts. Fixation in pupils' design and technology work (11–16 years)', *Design and Technology Education: An International Journal* 12(1): 34–44.

OfSTED (2011) *Meeting technological challenges? Design and technology in schools*. London: HMSO.

Prensky, M. (2001) 'Digital natives, digital immigrants, part 1', *On the Horizon* 9(5): 1–6.

Qualifications and Curriculum Authority (QCA) (2007) *The National Curriculum: statutory requirements for Key Stages 3 and 4, 2007*, London: Qualifications and Curriculum Authority.

Sandholtz, J.H., Ringstaff, C. and Dwyer, D. (1997) *Teaching with technology: creating student-centered classrooms*, New York: Teachers College Press.

Scrimshaw, P. (1997) 'Computers and the teacher's role' in Somekh, B. and Davis, N. (eds) *Using information technology effectively in teaching and learning*, London: Routledge.

Steeg, T. (2008) 'Makers, hackers and fabbers: what is the future for D&T?' in Norman, E.W.L. and Spendlove, D. (eds) The Design and Technology Association International Research Conference, Wellesbourne: Design and Technology Association.

Winn, D. (2012) *CAD and creativity at Key Stage 3: towards a new pedagogy*. Unpublished PhD thesis, Open University.

Chapter 11

STEM: opportunity, opposition or just good design and technology?

Tony Cowell

Introduction

In the centre of north Norfolk, cut into the wall of a parish church, is a collection of curved and straight lines. The church concerned is part of the remains of Binham Priory, the west front of which contains what is believed to be one of the earliest examples of Gothic tracery in England. The lines, unsurprisingly, as they are difficult to see, remained undetected for years until their discovery and mapping by the Norfolk Church Graffiti Survey in 2010 (Champion 2010). Champion describes the science that found the lines and explains they are now thought not to be graffiti, but the markings of the master mason who built the tracery, effectively a set of working drawings etched on the wall of the church. At this time in history, the mason would have been a craftsman, a maker of the highest quality, expertly skilled and accomplished. But was this mason a scientist, a technologist, an engineer or a mathematician? Did the concept of anything like STEM even exist in the early 1200s? Whatever description we use, this was undoubtedly someone able to apply the mathematical and scientific knowledge of the day, a true polymath, someone who could combine an understanding of all these areas of study to realise an innovative design of awe-inspiring beauty using the cutting-edge technology of the day. Several centuries on, this triumph lies in semi-ruin, a victim of the political and ideological change of the distant past, a reflection of how even the things society holds at the core of its values can soon become the focus for neglect and destruction.

This chapter explores the debate around design and technology (D&T) education and STEM (science, technology, engineering and mathematics). In several countries around the world, particularly America, STEM has been discussed for some time and is becoming increasingly important in educational discourse. Here, using the UK curriculum and context as the focus, I ask: is D&T a perfect vehicle to provide meaningful application of STEM in the curriculum and thereby to help educate the polymaths of the future? Or is STEM threatening to overwhelm traditional D&T activity, skewing learning opportunities solely toward technological and engineering contexts? These questions will be a thread running through the discussion, but STEM debate is a much more complex tapestry. The discussion begins with technology, and whether that means design and technology, then considers science and mathematics, eventually moving to engineering, a subject perhaps less easily defined in the UK curriculum. This involves shifting the perspective of the debate to consider the interests of industry, including the part design may play in this, pausing to reflect if neglect and political change could silence the STEM voice as effectively as the voices of those who built Binham Priory were silenced in the sixteenth century.

What do we mean by STEM?

STEM, which has now been part of the education vocabulary for over a decade, can be very simply explained as an acronym for science, technology, engineering and mathematics, a loose grouping of subjects. However a key figure in the development of STEM, Professor Sir John Holman, cautions against simple interpretations. Holman (2010: 6) warns: '[L]ike many acronyms, STEM means different things to different people – even to people in the same country.' He goes on to raise a vital point when considering STEM: 'Science, technology, engineering and mathematics look very different depending on whether they are viewed from inside or outside the classroom' (ibid: 6). This highlights our first important understanding: that the STEM subjects and their relationships to one another have different contexts, both an educational context and the wider one within society, particularly business and industry. Hence, the initial context for discussing STEM in this chapter will be from within the field of education.

Once this distinction is made we approach other issues. Pitt (2009: 37), when discussing STEM within the context of education for sustainable development, makes the point: 'At one end it is seen as a pre-vocational learning or even training to encourage students to pursue science and maths in particular en route to professional work in engineering and technology.' Holman and Pitt both indicate that STEM education can be seen as career preparation but they also present alternative perspectives as Pitt elaborates: 'Conversely, STEM can be seen as an entitlement to learn in a different way, in which the boundaries between the component subjects of STEM become blurred' (ibid: 37). However, the relationships between the subjects can also create debate and even tension. One study, involving science and technology teams working together, demonstrates that the interdisciplinary approach is not at all easy. In highlighting the uneasy relationships, it concludes: '[T]o overcome the professional socialisation that has led science and D&T teachers to be as they are will be difficult' (Lewis, Barlex and Chapman 2007: 57).

T for technology?

The working drawings of a medieval monastery may seem an unusual image on which to start a debate around the relationship between D&T and the STEM subjects. However, the drawings emphasise that without the effective synthesis and application of STEM knowledge, human beings would not have had such a huge impact on the environments in which they reside. When we make things we have a reason, a need; the concept of designing for a need has been with us for a long time in D&T education. That need, and the effectiveness of the solutions designed and manufactured to satisfy it, drives our species forward. Throughout history humankind has applied science, technology, engineering and mathematics to facilitate our evolution into one of the most effective species on earth, able to adapt to almost any environment. In his book *The artificial ape*, Timothy Taylor argues that technology has not been a product of our evolution but rather: 'Instead, the technology evolved us' (Taylor 2010: 9). He uses a range of examples to discuss this idea but what is most interesting from a D&T perspective is his focus on food and our development of cooking; textiles, in the form of our need to carry our young; and the tools we use to manipulate materials. This is a clear echo of three of the material areas recognised as being within design and technology – food, textiles and resistant materials. The idea that technology has evolved humans is, for some, a challenging concept but it does point to an interesting debate around the nature of technology in society and how it has changed. In exploring technology as a concept Taylor draws attention to the work of Tim Ingold, discussing the root of the word technology:

> Etymologically 'art' is derived from the Latin *artem* or *ars*, while 'technology' was formed upon the stem of a term of classical Greek origin, namely *tekhne*. Originally, *ars* and *tekhne* meant much the same thing, namely skill of the kind associated with craftsmanship.
>
> (Ingold 2001: 17)

Whoever the mason was who created the window at Binham, it is very probable that he understood the art, craft, design, manufacture, science, mathematics, engineering and technology of the time. How do we describe such a person now? This was someone whom James Dyson would perhaps term a 'hands and brains person' (Dyson 2010: 20), but how do we help develop this ability to synthesise technological concepts in the learners of today? Must we first define what we mean by technology in education? Western culture is not alone in this debate regarding the meaning of the word 'technology' and its application in education. Matsuda, in de Vries and Mottier (2006: 233), discusses a similar issue in Japanese technology education. Design and technology as a subject in schools has strong associations with skill and craftsmanship; could this then be what we mean by the term 'technology' in STEM?

Considering the STEM acronym has been used in education for 10 years or more, it does seem that D&T has taken a considerable amount of time to fully vocalise the contribution it can make to the STEM subjects. Kimbell (2010: 6) also drew attention to the fact that 'we should be clear that technology means DESIGN and technology, though STEM doesn't appear to like the *design* word.' Kimbell was discussing a report entitled 'Lengthening ladders, shortening snakes' (University of Warwick 2009), which is focused on STEM careers awareness. The report suggests an alternative perspective: 'How to challenge the widely held view in schools that technology is effectively "computing"?' (ibid: 10). This apparent negation of design and technology must be considered. Information and communications technology (ICT) has always had a close relationship with D&T and, as information technology (IT), it was part of technology in the original National Curriculum (DES 1990). Perhaps few at that time could have foreseen the huge impact the digital revolution would have on our understanding of 'technology'. Has the digital revolution changed our understanding of the word technology and might it have helped create the view raised in the Warwick report: that technology and computing are synonymous?

Yet there are also many who view ICT as a tool for learning, something to be embraced and taught across the curriculum, rather than a subject to be studied in isolation. Hague and Williamson (2009: 3), for example, noted that: 'Just as school subjects provide young people with the knowledge and skills to make sense of their world ... education should also supply the skills and knowledge to make sense of this digital media world.' Design and technology teaching may embrace this digital culture, with CAD/CAM activity, animation, video and other technologies exploring virtual design and manufacture, but in doing this it may tilt pupils' experience away from kinaesthetic activity and if, as Sigman (2008: 15) claims 'Human beings have an innate need to see, touch, taste, feel, and hear (experience) the features of any new object in order to understand it better', is digital experience only one part of a full D&T entitlement?

How then does D&T stake a strong claim to represent the T in STEM, alongside and perhaps incorporating ICT? Is there opportunity across the range of food technology, materials technology, textiles technology, electronic and communications technology? These areas within D&T each have a different relationship to STEM, yet all clearly have strong links with science and mathematics, for example electronics and control technology has direct and very obvious subject knowledge related to physics. In other areas, it could be argued that the STEM content is not immediately explicit: textiles for instance which is often perceived as a very visual medium, close to art and focused on fashion. Yet textile design and manufacture are being revolutionised by developments in science, technology and engineering: 'Known as electronic textiles, the new generation of fabrics are fibrous substrates into which microelectronic components and connectors have been seamlessly integrated.' (Quinn 2010: 10). Technical textiles and new materials such as QTC (quantum tunnelling compound) are changing the way we think of textiles as a design medium. The examples Quinn (2010) outlines show a huge range of scientific developments in textiles and suggest how textiles can be used not only for designing but also for engineering.

In contrast, other curriculum developments have seriously undermined the value that D&T brings to STEM education. In food technology, for example, the Licence to Cook initiative had a major impact on what was taught in the food technology curriculum. Rutland (2008) identified the concerns in her paper, 'Licence to Cook: the death knell for food technology?'. Rutland defines food technology in a manner that clearly highlights its connection to STEM, stating: 'Essentially, food technology covers an understanding of the nutritional, physical, chemical and sensory properties of food materials and how to apply this knowledge in developing food products' (Rutland 2008: 64).

It should be clear, then, that design and technology, represented by the T in STEM, can be shown to have valid and worthwhile links to the other STEM subjects from all its material areas. It could also be argued that good D&T teaching, irrespective of material, *is* applied STEM activity.

The relationship between T, S and M

There is a perception in D&T education that STEM is much more concerned with science and mathematics than it is with technology and engineering. Is this because science and mathematics are given more importance in the school curriculum? In the UK, as in many other countries, science and mathematics are considered 'core subjects' unlike technology and engineering, which are often optional.

Yet Lewis, Barlex and Chapman (2007: 37) noted that: 'Interaction between science and design and technology (D&T) in UK secondary education was recognised as being desirable as long ago as the 1960s.' There have been many attempts to develop this relationship and many texts discussing it. Barlex and Pitt (2002: 190) suggest possible ways forward for the relationship based on co-ordination and collaboration but they express a 'vigorous opposition to any notion of an integration of the two subjects'. Is it possible that in the development of a 'subject-focused' curriculum, we have compartmentalised learning and provided what Barlex and Pitt describe as a 'demarcationist view of the relationship between science and technology' (ibid: 186)?

Developments and revisions in the UK National Curriculum were accompanied by reflection on the relationship between subjects and on the teaching methods used in both science and technology. Banks (2006) takes an interesting perspective, focusing on the approaches to science and technology and the types of knowledge with which each is concerned. He makes a simple but important point: '[T]echnology should be more than just acquiring procedural skills and science should be more than the memorising of facts and the manipulation of abstract concepts' (Banks 2006: 198). Banks raises several important issues, one of which is that technology can learn from the values-led approach of the science curriculum development project 'Science Technology and Society', which is support for Barlex and Pitt's (2002) idea of science and D&T sharing good practice and co-ordinated approaches. Banks suggests that D&T could expand its use of group work activities and introduce debate around 'values' in design and technology. Since Banks wrote his book some of this can be seen in schools, but it is not yet a common experience.

STEM, then, is not only about sharing common learning through application but also about sharing practice across subject disciplines. However, bringing this approach to school has never been easy, as Lewis, Barlex and Chapman (2007: 56) conclude: '[T]his case study has encapsulated problems associated with teachers' intense ownership of intellectual capacity and particularly how they fail to understand and value each other's possible contributions.' In an effort to develop teacher subject knowledge and foster a co-ordinated approach, the Design and Technology Association developed a range of courses entitled 'Teaching Tomorrow's Designers and Technologists Today' (D&T News 2010: 17). The courses, which are offered through the network of science learning centres, are supported by science teachers and STEM ambassadors (STEM employees from industry who contextualise the content). The science content in these courses is explicit and delivered in a science environment, but the context is set in D&T activity. While this is helpful to some extent, it does raise some questions. Teachers leave the course with an electronics kit-based project exemplar but if this is then taught following only the procedures learnt, if the knowledge and understanding is not rooted in an effective design context that allows exploration of creative solutions, how can it claim to teach anything different from a science practical lesson. The courses also seem to be based on D&T teachers learning about science, if there really is to be co-ordination and collaboration then science teachers also need to learn about D&T.

There may also be interdisciplinary misunderstandings in the relationship between D&T and mathematics. A study by Nardi and Steward (2003), focusing on disaffection in Y9 pupils (aged 13 years) in mathematics, raises interesting issues around pupils' disaffection with mathematics teaching. The study focuses on a 'group of students whose disaffection is expressed in a tacit, non-disruptive manner, namely as disengagement and invisibility' Nardi and Steward (2003: 345). It is worth noting that the students were from middle ability groups and were not students typically seen as extreme in their behaviour. The study notes that: '[C]learly, these students view mathematics as an irrelevant and boring subject, the learning of which offers no opportunity for activity' (ibid: 351). It goes on to discuss issues around practical activity as potential motivation and comments on research which 'has repeatedly attributed student alienation from mathematics to its abstract nature and its heavy and compressed symbolic representation'; this seems to indicate mathematics teaching needs practical applications.

Many in D&T education are well aware of the positive benefits to students of practical activity, both in terms of behaviour and cognitive development (Sigman 2008). Yet the Nardi and Steward study, when discussing its own findings, shows the students do not engage with or see the point of the specific practical tasks, which were based on decorating a table or making a fence. From a D&T perspective, it seems obvious that the problem was the use of an inappropriate closed context for the application of the learning. Design and technology educators have given consideration to the way in which we can engage learners, and develop their creativity, through the way we present projects and define or inhibit their solutions. McLellan and Nicholl (2009: 89) suggest that: 'Similar arguments may apply to other areas of the curriculum where students are expected to generate creative ideas.' It seems that mathematics educators could learn something from D&T.

What opportunity is there to develop new, challenging and exciting STEM activity in a curriculum that focuses purely on academic subjects if the co-ordination Barlex and Pitt (2002) call for in science education does not also extend to mathematics? Is it time we to consider how we can begin 'taking the lead in blurring the boundaries' (Pitt 2009: 45).

E for engineering

Harrison (2011: 17) noted that: 'Engineering, the "E" in STEM, is seldom taught as a distinct curriculum subject in English and Welsh schools to pupils under the age of 14 years', but from the very early days of the National Curriculum the engineering industry has commented on D&T education and its ability to prepare pupils for work in the engineering sector. The initial D&T National Curriculum was not well received by many, as Wakefield and Owen-Jackson report in Chapter 1, including representatives from engineering organisations. Lewis, Barlex and Chapman (2007: 39) point to a briefing for the Engineering Council which begins: 'Technology in the National Curriculum is a mess' (Smithers and Robinson 1992: 5). Despite the criticism the briefing document warns of a past where 'English education was (and perhaps still is) steeped in the classical tradition which saw intelligence as distinct from the lower faculties used in practical activities' (ibid: 19). In 2001 Sir John Roberts was commissioned by the government, as part of its strategy for improving the UK's productivity and innovation performance, to produce a report on the supply of people with skills in science, technology, engineering and mathematics. This 'stemmed from the Government's concern that the supply of high quality scientists and engineers should not constrain the UK's future research and development (R&D) and innovation performance' (Roberts 2002: 1). The review considers D&T and its contribution to science, engineering and technology (SET – the forerunner to STEM). It discusses issues that, at the time, it was considered needed to be resolved, such as 'poor environments in which science, and design and technology practicals are taught' (ibid: 3) and 'the ability of these subjects' courses to inspire and interest pupils, particularly girls' (ibid: 3). It also highlights the need to train D&T teachers. The report presents data that may be considered as the basis for the focus on science and mathematics in the following years. The data include a graph of engineering and technology A level entries for the period 1995–2000 mapped against entries to engineering and technology degree courses by the same cohort. This shows a marked contrast between the sharp increase in D&T entries while, at the same time, a decrease in entry to degree programmes in engineering and technology. The report draws the conclusion: 'This seeming disparity arises, in part, because other subjects such as mathematics and physics – which are in decline at A-level – are also very important in preparing students to study engineering in higher education' (ibid: 24). Did this then present to some a rationale to place more resource into science and mathematics?

This discussion also raises the question of whether or not universities and industry accept D&T A level as an appropriate qualification for entry to engineering and technology degrees. Could this be seen as industry presenting a strong case to influence education? It is easy to see where this suspicion comes from. The engineering sector continued to be involved with the development of the curriculum, especially after the publication of the Tomlinson Report (2004). It is clear that, by 2006, STEM had moved on apace and yet still D&T featured little in its remit, as Barlex (2008: 16) noted: 'The original Science, Technology, Engineering and Mathematics (STEM) Report in 2006 virtually ignored our subject.' However, SEMTA, the sector skills council for engineering and associated industries, produced several reports highlighting the potential skills gap as trained engineers retire (SEMTA 2010). They present a 'need' for STEM-focused entrants and worked hard in influencing the STEM careers initiatives such as Future Morph. These careers initiatives clearly present STEM as a driver in the economy: 'Science, technology, engineering and mathematics (STEM) are critical to the UK's future prosperity' (University of Warwick 2009: 3).

D for design

Although there is not a D in STEM, Technology has to be seen to embrace 'design and technology' so consideration of design in relation to STEM is important. In 2005 Sir George Cox was commissioned by the UK government to produce a report on 'Creativity in business'. Cox places design and creativity in a powerful position to link with business but goes on to state: 'It is common for those in business to see creativity and the related area of design as largely concerned with aesthetic considerations such as style and appearance' (Cox 2005: 3). This report was the key factor that led the Design Council to set up a multidisciplinary design network. A subsequent Design Council report emphasises the importance of a multidisciplinary approach to design and is clear about the role of design in STEM subjects: 'Introducing design thinking tools to science and technology students is one way in which we can improve the UK's ability to capitalise on emerging technologies' (Design Council 2010: 21). Evidence from around the world suggests that STEM needs to encompass the methodologies and opportunities for developing creativity and innovation inherent in art and design. Other reports also consider design and STEM: 'To a large extent, the STEM agenda has (also) ignored its silent D (design)' (Dyson 2010: 21). Perhaps the design industry is now becoming as interested as the science, engineering and technological industries in the STEM debate and STEM education.

The introduction of university technical colleges (UTCs – a new type of school that may now be set up in England) suggest that the government favours the classical tradition mentioned by Smithers and Robinson (1992). The government has defined UTCs as 'academies for 14–19-year-olds. They focus on providing technical education that meets the needs of modern employers. They offer technical courses and work-related learning, combined with academic studies' (DFE 2011). If UTCs only offer STEM-focused 'vocational' design and technology activity where will this leave opportunity for design and creativity?

This takes us back to the start of the chapter and the discussion regarding the definition of technology. If design has such a key part to play in STEM, then perhaps design and technology can stake a strong claim to be the T of STEM in schools?

Conclusion

The curricula future is still uncertain and it is interesting to consider the range of interest groups trying to influence the curriculum in a specific direction, either to provide the PhD graduates of the future or the vocational apprentices of industry (Banks 2006: 208). This academic–vocational divide is still a key debate (see Chapter 6) and recent proposals in England seem to be emphasising the academic. The White Paper (DfE 2010: 11) seeks to: 'Introduce the English Baccalaureate to encourage schools to offer a broad set of academic subjects to age 16, whether or not students then go down an academic or vocational route.' Some would argue that a baccalaureate curriculum will prevent the interest groups directing students at age 14 into specific routes, but it is worth noting that the proposals for the English baccalaureate curriculum include science and mathematics but has no mention of technology or engineering. There is also no mention of STEM as a group of subjects. The baccalaureate seems to point to an academic route for all, with academic or vocational options later.

Does this ring a death knell for STEM as an integrated learning experience? Will this clear divide mean that more academic students do not have the full opportunity to apply their learning in a 'real-world' context? If we place D&T firmly in a STEM context could we find ourselves at the whim of political change driven by the perceived need or otherwise for STEM students in higher education and industry?

Travelling to Binham to search for medieval graffiti produced at a time in history not long after one that 'was sunk in a Dark Age of dynastic squabbling, pagan aggression and superstitious poverty' (Gove 2011, para 12) seemed to draw a parallel with the current state of the STEM debate. Does Gove's speech, from which this description of dark age western society is taken, indicate a move away from STEM back to the science and mathematics focus of the past? Does the UK government policy of 2011 reflect the view that D&T activity is a craft-based alternative to academic study, unable to contribute to a knowledge-based economy? These are questions that are still to be answered – yet in this same speech Gove states: 'If we are to keep pace with our competitors, we need fundamental, radical reform in the curriculum, in teaching, and in the way we use technology in the classroom' (Gove 2011). Design and technology is rising to that challenge and proving its value through initiatives such as STEM training, courses in e-textiles,

STEM clubs and activities in schools, and it must continue to do so. Barlex (2008), in an article titled 'Can we still be cheerful?', describes opportunities for D&T with STEM and concludes the pendulum was moving away from science and mathematics and towards D&T-focused technology and engineering. The introduction of the English baccalaureate and the speech by Michael Gove to the Royal Society might be taken to indicate that the pendulum swing may have ceased politically but it is to be hoped many schools and teachers will maintain the momentum and realise the value that D&T can contribute to scientific and mathematical understanding through the practical application of knowledge.

The chapter title questions whether STEM is an opportunity for D&T, to gain status and value by alignment with 'academic' subjects; in opposition to it, by seemingly taking over D&T and forcing it off the school curriculum. Or is it just good design and technology teaching that embraces cross-disciplinary subject matter and addresses it in a meaningful context which engages pupils and supports their learning? Design and technology might just be the perfect vehicle to navigate an experiential learning journey that really will help create the STEM polymaths of the twenty-first century. If D&T 'really is the integrator, the sense maker, the interest-provider that transforms arid and failed models of learning and brings them to life' (Kimbell 2010: 8) then beware society! For without meaningful D&T in the education system creating opportunities to continually apply science and mathematics in real contexts, how will there be anyone with the true understanding and innovative ability to join the dots and engineer a new, sustainable future for the world?

Questions

1   What opportunities do design and make activities create for STEM learning?
2   How important is design and creativity to STEM education?
3   How does design and technology ensure the wider community sees its contribution to STEM as more than systems and electronics?

References

Banks , F. (2006) 'Technology, design and society (TDS) versus science, technology and society (STS): learning some lessons' in Dakers, J.R. (ed.) *Defining technological literacy towards an epistemological framework*, Basingstoke: Palgrave Macmillan.
Barlex, D. (2008) 'STEM: can we still be cheerful?', *D&T News* 38.
Barlex, D. and Pitt, J. (2002) 'The relationship between design and technology and science' in Owen-Jackson, G. (ed.) *Teaching design and technology in secondary school*, London: RoutledgeFalmer.
Champion, M. (2010) 'Reading the writing on the wall – the Norfolk Medieval Graffiti Survey', *Current Archaeology* 256: 41.
Cox, G. (2005) *Review of creativity in business: building on the UK's strengths*, London: HM Treasury.

D&T News (2010) 'Dates for your diary – D&T conferences and events', *D&T News* 45.
DES (Department of Education and Science) (1990) *Technology in the National Curriculum*, London: Department of Education and Science and the Welsh Office.
Design Council (2010) *Multi-disciplinary design education in the UK. Report and recommendations from the Multi-Disciplinary Design Network*, London: Design Council.
de Vries, M. and Mottier, I. (eds) (2006) *International handbook of technology education*, Rotterdam: Sense.
DfE (Department for Education) (2010) *The importance of teaching – the schools white paper 2010*, London: TSO.
DfE Department for Education) (2011) *What are UTCs?*, http://www.education.gov.uk/schools/leadership/ typesofschools/technical/a00198954/utcs (accessed 29 October 2011).
Dyson, J. (2010) *Ingenious Britain – making the UK the leading high tech exporter in Europe*, London: Dyson Foundation.
Gove, M. (2011) *Michael Gove speaks to the Royal Academy on maths and science*, http://www.education.gov.uk/inthenews/speeches/a00191729/ michael-gove-speaks-to-the-royal-society-on-maths-and-science (accessed 29 October 2011).
Hague, C. and Williamson, B. (2009) *Digital participation, digital literacy, and school subjects – a review of the policies, literature and evidence*, http://archive.futurelab.org.uk/resources/documents/lit_reviews/DigitalParticipation.pdf (accessed 28 October 2011).
Harrison, M. (2011) 'Supporting the T and E in STEM: 2004–2010', *Design and Technology Education: An International Journal* 16(1): 17–25.
Holman, J. (2010) 'Forward', *Design and Technology Education: An International Journal* 16(1): 6.
Ingold, T. (2001) 'Beyond art and technology: the anthropology of skill' in Schiffer, M.B. (ed.) *Anthropological perspectives on technology*, Albuquerque: University of New Mexico Press.
Kimbell, R. (2010) 'Handle with care', *Design and Technology Education: An International Journal* 16(1): 7–8.
Lewis, T., Barlex, D. and Chapman, C. (2007) 'Investigating interaction between science and design and technology (D&T) in the secondary school – a case study approach', *Research in Science & Technological Education* 25(1): 37–58.
McLellan, R. and Nicholl, B. (2009) 'If I was going to design a chair, the last thing I would look at is a chair. Product analysis and the causes of fixation in students' design work 11–16 years', *International Journal of Technology and Design Education*, Springer Online First,™ 17 December.
Nardi, E. and Steward, S. (2003) 'Is mathematics T.I.R.E.D? A profile of quiet disaffection in the secondary mathematics classroom', *British Educational Research Journal* 29(3): 345–366.
Pitt, J. (2009) 'Blurring the boundaries', *Design and Technology Education: An International Journal* 14(1): 37–48.
Quinn, B. (2010) *Textile futures: fashion design and technology*, Oxford: Berg.
Roberts, G. (2002) *SET for success – the supply of people with science, technology, engineering and mathematics skills review*, London: HM Treasury.

Rutland, M. (2008) 'Licence to Cook: the death knell for food technology?' in Norman, E.W.L. and Spendlove, D. (eds) The Design and Technology Association International Research Conference, Wellesbourne: Design and Technology Association, 2–4 July.

SEMTA (February 2010) *England report – skills and the future of engineering in England*, London: Sector Skills Council for Science, Engineering and Manufacturing Technologies.

Sigman, A. (2008) *Practically minded – the benefits and mechanisms associated with a craft-cased curriculum*, Stroud: Ruskin Mill Educational Trust.

Smithers, A. and Robinson, P. (1992) *Technology on the National Ccurriculum – getting it right*, London: Engineering Council.

Taylor, T. (2010). *The artificial ape – how technology changed the course of human evolution*, London: Palgrave Macmillan.

Tomlinson, M. (2004) *14–19 curriculum and qualifications reform – final report of the Working Group on 14–19 Reform*, London: DfES Publications.

University of Warwick (2009) 'Lengthening ladders, shortening snakes – embedding stem careers awareness in secondary schools', London: Centre for Education and Industry; http://www2.warwick.ac.uk/ fac/soc/cei/ (accessed 29 October 2011).

Recommended reading

Design and Technology Education: *An International Journal* 16(1) – this entire journal is devoted to discussions on STEM and considers aspects detail. It also points to myriad sources for further reading.

Dyson, J. (2010) *Ingenious Britain*, Dyson Foundation – this report asks some very interesting questions about the future of STEM education.

Kingsford, P.W. (1964) *Engineers, inventors and workers*, Arnold – an old school text, but invaluable reading regarding the history of engineering and the polymaths responsible for the industrial revolution.

Taylor, T. (2010) *The artificial ape – how technology changed the course of human evolution*, Palgrave Macmillan – although not STEM related this is a thought-provoking discussion on the impact of technology on humankind.

Part IV

Debates about teaching design and technology

Chapter 12

The (continuing) gender debate

Dawne Bell, Chris Hughes and Gwyneth Owen-Jackson

Introduction

There has long been a gender divide in the subject areas that make up design and technology (D&T). Historically, boys would invariably follow a workshop-based curriculum based on metal and wood manufacturing skills, while girls would follow a curriculum preparing them for domestic life through subjects such as cookery and needlework. In the UK, the Sex Discrimination Act 1975 made such gender divisions illegal and these subjects were opened up to pupil choice and pupils still chose along gender lines. The introduction of design and technology, in 1990, also introduced the requirement for girls and boys to study all areas of D&T in lower secondary school, giving them a broader curriculum than previously. In upper secondary school, however, pupils chose one area to study to examination level and the gender divide remained, with the majority of boys choosing resistant materials or electronics and the majority of girls choosing food technology or textiles. Why is this, and what – if anything – should we do about it?

This chapter, first, discusses what we mean by gender and its link to learner identity. There are then several debates around gender and D&T, we consider what we mean by 'technology' and whether it is our definition of it that is creating gendered perceptions. We look at the gender differences in learning about D&T and the implications of this. Finally, we consider possible ways to address the continuing gender concerns in design and technology.

Gender and learner identity

Your sex is physically determined; you are born female or male. Gender, however, describes those characteristics of men and women that are socially constructed and formed though the interaction with the dominant norms and values of society, socialisation in childhood, peer pressure at different stages in one's life and perceived societal roles. Gender is ascribed from an early age, in all societies and cultures. In the UK, for example, a baby dressed in pink clothing is taken to be female and one in blue clothing is assumed to be male and, although the colours vary in different countries the expectations remain the same. Gender expectations are further demonstrated through the clothes and toys bought for children, with girls being given dolls and tea-sets and boys building blocks and train sets. At later stages, these gender influences can impact on subjects studied and career choices and so have a determining influence on an individual's life.

There are further social impacts, in that male gender influences are generally perceived more positively than female ones. For example, girls can dress in jeans without comment but boys are unlikely to be dressed in skirts. Similarly, girls demonstrating male characteristics and enjoying boys' games are called 'tomboys' and accepted, but boys demonstrating female characteristics and playing girls' games are called 'cissies' and are less acceptable. Although these barriers are beginning to be broken down, they still remain. In some cultures, the gender divide is strongly reinforced and straying beyond it is frowned on.

These gender influences are important within D&T. First, the areas perceived as the male areas of the subject – resistant materials, electronics, systems and control, engineering – tend to be held in higher esteem than the areas perceived as the female areas – food and textiles. This may have implications for pupils' subject choices. Second, gender also seems to have significant influence on the way girls and boys relate to technology and identify the technological aspects of their lives. The ways in which boys and girls engage with technological projects in D&T have to be considered in the context of powerful cultural forces that shape and describe the way they think, act and perform. Factors such as advertising, the gender assignment of children's toys, perceptions of 'masculine' and 'feminine' roles and appropriate jobs have bearings on how boys and girls begin to form their ideas about the gendered roles they play in society, ideas that often override the gender neutral ideals sought within the D&T classroom.

Gender is also linked to the individual learner's identity. Learner identity has been described as 'the tool that enables the construction of meaning about oneself as a learner' (Falsafi 2010: 6). It encompasses other sociocultural identities we each construct for ourselves, such as our class and role identity and is likely to impact on the approach taken to learning and participation in learning, including subject choices. However, how much of this is predetermined and how much is socially constructed? The phrase 'nature versus nurture', coined originally by Galton (1869), alludes to the debate around how much the environment can 'strengthen or thwart' pre-existing tendencies. That is to say, are we born with a tendency to favour one thing over another or is it the nurturing processes to which we are subjected that influences the choices we take later in life?

Browne (2004) explores the issue of gender equity in the early years curriculum and her studies of children at play give rise to evidence claims of significant differences in the way boys and girls play that lead to, and account for, the gender differences in the skills set they develop. Browne and Ross (1990) observed that when presented with construction materials boys build complex structures, helping them to develop manipulation, exploration and visual-spatial skills

(Newcombe 1982). Presented with the same challenge, girls built less complex constructions and went on to incorporate these into social play.

In their work, Caleb (2000) and Sanders et al. (1997) note that generally boys' toys – by which they mean toys that are considered appropriate for boys to play with – are more technological and are designed to encourage construction and manipulation. Toys bought for girls, in contrast, tend to promote the development of interpersonal social skills. However, others, including Miller (1987) would argue that when presented with gender-neutral toys, such as those in 'junk play' (Stagnitti et al. 1997), boys and girls play with such toys in different ways because of their previous exposure to other gender-stereotyped toys. Sanders et al. (1997) found that girls who are not exposed to 'boys' toys' are less likely to develop an interest in technologically related subjects later on.

In secondary schools, researchers (Kimbell et al. 1991; Murphy 2006) found differences in the way that boys and girls work and respond to tasks in D&T. Girls tended to focus their context on people and social needs and considered these in their design work, while boys focused on the product and its technical aspects. These research findings do not provide evidence of whether the differences between boys and girls are innate or socially constructed, but they do provide evidence that boys and girls respond differently to tasks in D&T.

What do we mean by 'technology'?

The word 'technology' derives from the Greek, *technologia*, meaning the study of art, craft or skill but put it into any search engine and you will be given links to information about computer technology, space technology, energy, cars. As Wajcman (1993: 137) notes: 'The very definition of technology ... has a male bias.' Despite social and technological changes, and our understanding of technology growing, developing and changing over the years, in 2009 the UPDATE project reported that 'perceptions about technology are still strongly aligned with the concept of masculinity' (Dakers and Dow 2009). There are now different 'technologies' and the word can be used to refer to the machines and tools used or to the processes involved. This partly accounts for the differences in technology curricula around the world (as described by Banks and Williams in Chapter 3) and for the lack of understanding around 'design and technology' in schools today. Design and technology, in England, covers the study of resistant materials (wood, metal, plastic), electronics, systems and control, food and textiles. Yet D&T, or technology, is still often perceived as a 'masculine' subject; why is this?

There are several possible reasons. The original National Curriculum drew together disparate subjects, including art, business studies, craft, design and technology, home economics and information technology, to create 'technology' as a new discipline. At the time there was great debate over whether or not home economics should be included but in the final outcome it was. After several curriculum reviews, art, business studies and information technology became separate subjects and 'design and technology' became the study of what had previously been craft, design and technology (CDT) and home economics. As 'technology' had previously been a recognised part of CDT it was considered by many that D&T was an updated version of CDT. This gives one reason why D&T is perceived as masculine, although it is now an alliance of the different subject areas; D&T is often used as shorthand for the areas of resistant materials and electronics with food and textiles studies referred to as separate subject areas.

Resistant materials and electronics are the subjects previously taught only to boys, they are taught in workshop areas, often by male teachers. This leads to the perception that this is an area for 'boys' and many girls do not feel comfortable in the environment. Examination entry data for 2009 show that 9.2 per cent of girls took an examination in resistant materials compared to 39.9 per cent of boys. In electronic products, only 0.8 per cent of girls entered the examination and seven per cent of boys (Kirkup et al. 2010). However, this may be subject to change with the growing use of computer-based machinery and the increase in the number of female teachers in this area (data are not collected so statistics are not available, but informal evidence from teacher training institutions indicates growing numbers of female student teachers in resistant materials and electronics).

There is also a (western) feminist argument that because food and textiles are considered domestic, and so predominantly female, they have a lower status than resistant materials and electronics (see Wajcman 1993; Cockburn and Furst Dilic 1994). This again serves to leave food and textiles as the junior partner in D&T. The impact of this can be seen in the various initiatives set up to address the gender imbalance (see the section 'Addressing the issues'). There were initiatives to encourage girls to study 'technology' but none to encourage boys to study textiles or food. This is not only due to the gendered perception of these two areas, but also to the lack of awareness of the 'technology' involved in food and textiles studies.

So why is food technology not considered to be technology? This is partly due to its historical association with domestic food production and home economics and, as discussed in Chapter 8, the difficulty that some teachers found in embracing the philosophy of D&T. Food technology, in schools, has struggled to identify itself with D&T working practices and to develop the industrial, technological approach that is valued. Yet such an approach is possible. Food technology can encompass biotechnology, technological production processes and food science and technology, but often in schools does not (Rutland and Owen-Jackson 2012).

Despite the huge social and public interest in food, with extensive television and media exposure and a plethora of celebrity chefs, the majority of whom are male, this area remains a female domain. Boys are required to study food technology in lower secondary school but when making their examination choices at upper secondary level, many choose not to continue. In 2009, 15 per cent of boys entered the examination for food technology compared to 33 per cent of girls (Kirkup et al. 2010). This, again, is linked to perceptions of it as a 'female' subject, its historical links with the home and the high number of female teachers. There are increases in the number of men training to teach food studies, but this remains small compared to the number of women. The introduction of catering as an examination option has seen an increase in the number of boys studying the subject, but this is regarded as a practical, vocational subject and not a technological one.

Textiles technology is also predominantly a female-based discipline, with even smaller numbers of male teachers and boys electing to study it at examination level. At examination level, this is almost 100 per cent female entry with only 0.8 per cent of entrants being boys (Kirkup et al. 2010). This is less easy to explain as the working practices in textiles technology are similar to those in resistant materials and electronics and there is increasing use of machinery, which should appeal to boys. There is also increasing emphasis in the industrial and technical use of textiles, for example in sportswear and protection clothing, and an increasing use of electronics within textiles. There is no apparent reason, other than its close connection with female, domestic activity, for boys choosing not to study textiles technology.

The teaching of D&T may also be partly responsible for its masculine identity. Wajcman (1993: 152) reports that studies found: '[T]eachers behave differently to girls and boys, they speak to them differently, they require different responses and different behaviour from them', a finding supported by Murphy (2006). In their study, 'de-constructing masculinity' Dakers et al. (2009) explore the perceptions of secondary school age pupils in relation to technology education and pedagogy. They found that the teaching is perceived to have masculine traits, which curtails girls' interest in studying it.

Despite girls' reluctance to study 'technology', however, when they do the evidence is that they do well.

Gender and attainment

Historically, examination data (Harding 1997) show that within D&T both genders tend to choose their respective traditional subject areas, with boys opting for workshop-based subjects and girls favouring food and textiles. In 1982 only 1.6 per cent of GCE O level woodwork examination entrants were girls. In 1992 there were 21.5 per cent girls entering the examination for 'resistant materials', the contemporary equivalent of woodwork, while 13.8 per cent of the food technology examination entrants were boys.

From 2004 in England, D&T was no longer a compulsory examination subject but it continued to be a very popular option choice for both boys and girls. An analysis of examination statistics for the period 2008–2010 (Design and Technology Association (DATA) 2011) shows a steady decline in examination entries across all the D&T areas, but that more boys than girls choose to study D&T to examination level. However, further analysis of examination results illustrates that the girls who chose to study D&T outperformed their male counterparts, with 71.8 per cent of girls gaining grades A* to C compared to 55.1 per cent of boys (Kirkup et al. 2010). This 17 per cent 'gender attainment gap' between boys and girls is significant, and the largest in the National Curriculum subjects. But why do girls outperform boys?

It has not always been the case that girls have performed better in examinations and it is not unique to D&T, so why has it changed? While for individuals it could be due to a number of factors, including variance in maturation levels and socio-economic circumstances, many (Elwood 1999) cite the introduction of coursework-based tasks as the central attributing factor for the improvement in girls' achievements. In D&T, coursework has been the major contributor to examination grades and since coursework was introduced as part of the examination process, in the 1980s, girls have continued to do better. Shepherd (2010) argues that this is because boys tend to favour short, snappy 'instant reward' style work and so excel in traditional written examination styles (Measor 1983) while girls prefer to develop and refine their work. Research by examination boards (Curtis 2009) appears to substantiate these claims.

So, given the influences of gender and learner identity, the perception of technology as masculine and the evidence of attainment data, what could be done to address the gender imbalance?

Addressing the issues

The issues are to do with the perception of D&T as a masculine subject, the lower status of food and textiles within D&T and the gendered study of areas within D&T. There are several ways in which schools and teachers can address these issues, including:

- research initiatives, testing and trialling different approaches
- presenting an appropriate learning environment
- reconsidering the teaching of D&T
- single-gender provision.

Research projects and initiatives

Beginning in the 1980s concern over the low number of girls studying science- and technology-based subjects led to the development of a series of projects that sought to identify and implement potential solutions. The projects and initiatives were designed to explore aspects of gender bias within technology and science and to encourage the increased participation of girls. The mid-1980s saw the development of several projects, including Women in Science and Engineering (WISE) in 1984, Girls and Technology Education (GATE) (Harding and Grant 1984) and Girls into Science and Technology (GIST) (Kelly et al. 1984). GIST was a four-year action research project that sought to investigate the reasons for girls' underachievement in the areas of science and technology. It has to be said, however, that there was more emphasis on science than technology, but these projects did serve to raise the profile of technology and encourage girls to study resistant materials and electronics. In terms of the gender imbalance, it is interesting to note that there were no similar projects to encourage boys to study textiles and food. There is little research on this, but it is likely that encouraging girls into science and 'technology' was an economic response rather than a social or gender-based one.

Following the introduction of the National Curriculum in 1990, and despite the projects and initiatives just mentioned, it was soon noted that at examination level girls were continuing to opt for traditionally 'female' areas of food and textiles and boys for resistant materials and electronics. This gave rise to a second series of projects, which ran throughout the 1990s, designed to explore the phenomena and devise and implement potential solutions.

Building on earlier work, the Girls Entering Tomorrow's Science, Engineering and Technology (GETSET) (McIntyre and Woolnough 1996) initiative was supported through locally based science and technology regional organisations (SATROs). This initiative sought to encourage girls to study science and technology through classroom-based work, rewarding achievement through certification such as the CREativity in Science and Technology (CREST) awards and the opportunity to engage in regional and national competitions.

The 'Newley' Technology Initiatives (Withey 2000, 2003) introduced innovative observational work focused specifically on gender interaction during practical lessons in resistant materials. The findings, which are supported by other studies (Francis 2011), highlighted the dominance of boys within the workshop environment. Withey also investigated teacher perspectives and differing expectations of boys and girls and how these, and subsequent actions, impact on pupils' engagement and achievement, an issue also studied by Ivinson and Murphy (2007).

The Understanding and Providing a Developmental Approach to Technology Education (UPDATE) project (Dakers et al. 2009) was a Europe-funded project investigating why girls dropped out of technology and how teaching could be amended to encourage them to continue. They recommended that, as gender attitudes are formed early, parental influences and early years schooling is important and should be reconsidered. They also recommended reviewing the methods used to teach technology, introducing a case study or critical pedagogy approach as a way of making technology more appealing to girls. It is not clear what impact this had in Europe but there is little evidence of the recommendations being taken up in England.

Many of these organisations and initiatives have evolved and are now encompassed, and funded, under the Science, Technology, Engineering and Mathematics (STEM) agenda. STEM (see Chapter 11 for a discussion of this) is another economically driven initiative that aims to encourage boys, as well as girls, to study these subjects to a higher level. However, technology within the STEM agenda refers to resistant materials, engineering and electronics within D&T; there is little support in STEM for the continuing study of food or textiles technology.

The design and technology learning environment

Most D&T learning environments are based around the machinery, tools and equipment used in the teaching of the subject. It has been suggested (Withey 2000, 2003) that girls find such environments unfriendly, so consideration could be given to ways in which to make the environment more appealing. This might include making sure there are exhibitions and displays of work that will support and encourage students. Displays of women inventors and technologists can also be effective through serving to show girls positive role models (Hill, Corbett and St. Rose 2010). There could be examples of various applications of technology, which could be selected for their appeal to the different genders. Similarly, in the food and textiles areas there could be displays of male textile designers and chefs, together with other examples such as food scientists, food product developers, textile workers. There could be posters showing the various technical aspects of food production and textile applications.

In order to increase boys' engagement and attainment in D&T, the learning should make more use of 'technology' (OfSTED 2011). Ways in which this could be done include greater use of CAD/CAM equipment (see Chapter 10) and mobile technology for recording the development of work (see Chapter 14) and more use of smart and modern materials.

These suggestions appear simplistic but improving the learning environment can serve to encourage pupils to enjoy coming to the learning area and form a more positive attitude to the work they undertake.

Developing a 'gender-neutral' curriculum

If we accept that there are differences in the preferences of boys and girls, how can we accommodate these in what we teach in D&T? Gender-neutral teaching involves teaching the concepts of D&T through topics or projects that appeal equally to boys and girls or which can be interpreted differently by different individuals. Examples of such projects in resistant materials would be making clocks or games, timer devices in electronics, and in textiles, items such as mobile phone holders.

The way in which D&T is taught may also need to be reconsidered. Some years ago the Assessment of Performance Unit (APU) (Kimbell et al. 1991), focusing on assessment processes in D&T, found that girls and boys performed better in different types of activity. Their findings, although presented with caveats, suggest that for girls to do well more attention should be given to the structure of the learning activities as well as the content. They found that girls preferred more open tasks but more tightly structured activities, while boys preferred more closed tasks and less structured activities. The APU report suggested that teachers should give careful consideration to the context of tasks, the open/closed nature of them and the structure of learning activities so that they can, as far as possible, support the learning of both genders.

More recently, the UPDATE project (Dakers et al. 2009) suggested that the traditional 'transmission' model of teaching technology, developing the mastery of knowledge and skills, was not appealing to girls. They suggest that social constructivist, using a case study approach, or critical pedagogies approaches seem to engage girls more.

These research findings show how crucial the teacher is in determining the engagement and performance of girls and boys in D&T. Teachers make decisions about what to teach and how to teach it within the constraints of national curriculum requirements and these decisions can have important consequences.

Similarly, who the teacher is can have an impact. Historically, and up to as recently as the 1980s, there were very few female teachers of craft, design and technology and correspondingly few – if any – male teachers of food or textiles. Numbers are growing slowly but still remain low. This situation is echoed in teacher training institutions where there are more male lecturers specialising in resistant materials, engineering and electronics and more female lecturers in food and textiles, although again numbers are slowing changing.

Does this matter? Well, for pupils it reinforces the stereotypical view that many of them have, that resistant materials and electronics are male dominated and food and textiles are female dominated. It may dissuade some pupils from taking a subject if he or she is likely to be the only male or female in the classroom. How might this change if the gender of the teacher is atypical? This is now increasingly common in resistant materials and, to a lesser degree, in food, and pupils are beginning to see a breakdown in the gender divide. This is not yet having an impact on subject choices for examination, but it may be the beginning of a slow process of change.

Teachers can also make a difference if they challenge stereotypical assumptions in the classroom (OfSTED 2011). As this chapter has shown, from an early age girls and boys are bombarded with male and female stereotypical characteristics and this can manifest itself in their design work. Girls may choose pictures of teddy bears and flowers, and prefer pink and pastel colours, while boys may choose cars and sports images with a preference for darker colours. These choices are often not appropriate for the product but are unchallenged by the teacher, who could ask why a particular design or colour has been chosen and whether or not alternatives were considered.

Single-gender teaching

Secondary schools in England are mostly mixed comprehensive schools, although this is now changing, but some single-sex schools still exist and these schools tend to feature heavily at the top of the annual league tables (Hurst and Sugden 2010), which are a measure of how well pupils perform in examinations. Does this mean that single-sex teaching is advantageous? There is no firm evidence to prove that this is the case (Smithers 2006; Smithers and Robinson 2006) and the reasons for their success could be due to factors such as pupils' socioeconomic background, smaller class sizes and levels of funding often associated with such schools.

There have also been several studies into single-sex teaching within co-educational schools (Jackson 2002; Parker and Rennie 2002, Younger and Warrington 2006). None of these studies provides evidence that single-sex teaching influences pupils to choose 'non-traditional' examination subjects. However, we would not dismiss this strategy. There is evidence that boys and girls differ in their approaches to learning (Gurian and Stevens 2010) and that they approach D&T tasks in different ways (Kimbell et al. 1991). Boys tend to prefer to work alone, focusing on the physical aspects of structure, mechanism and the need to 'get it done', whereas girls tend to share knowledge, consider the social aspect of a task and may become preoccupied with detail and aesthetical appeal. By teaching these groups separately, the teaching could be better focused on the majority learning style.

Conclusion

Despite the best efforts of many over the years, comparatively little has been achieved in changing the gender divide in D&T (OfSTED 2011). Does this matter? Well, for girls there may be economic benefits of studying the male-dominated technologies, in terms of better job prospects and higher pay (Paechter 2007; Kirkup et al. 2010). There is still under-representation of females employed within areas that have historically been considered to be male domains; for example statistics show that, in the UK, 5 per cent of working women have careers in science, engineering or technology (SET) compared to 31 per cent men and that, although women account for 45 per cent of the workforce they represent only 12 per cent of the SET workforce (Kirkup et al. 2010). This has economic consequences for both the nation and for individuals.

There is also the social issue of the perceptions and status of 'women's work'. According to Crawford and Unger (2004) what appears to be male tends to be valued and what is female or feminine tends to be devalued. If male teachers can be persuaded to teach food and textiles and boys encouraged to study them to a higher level this may help to raise the status of these areas and so make them more appealing to more pupils who, in turn, will benefit from the broader education they receive.

Despite the continuing gender divide in D&T, the gender debate seems to have shifted to be more concerned with the underachievement of boys rather than the subject choices of girls (Younger and Warrington 2006, OfSTED 2011). There are some concerns (Skelton and Read 2006; Charlton et al. 2007) that this change of focus will have a detrimental effect on the education of girls, as research shows that boys continue to dominate classroom spaces, teacher time and material resources (Withey 2003; Francis 2011).

Design and technology is a subject concerned not only with developing pupils' conceptual knowledge but also their skills, for example of creativity, problem solving, team working, decision making, analysis and evaluation. This knowledge and these skills are important both for personal enrichment and to support pupils in their career choices. By perpetuating the outdated gender divide D&T is doing a disservice to girls and boys. The subject cannot overturn social and cultural values but it can help boys and girls to see their influences and encourage them to see that both women and men can contribute to developments in technology – whether by this we mean encouraging girls to become engineers or boys to become textiles artists.

Questions

1 Has your education been influenced by your gender? If so, in what way?
2 Have you observed any gender bias in teaching in technology classrooms?
3 Do you think this is an important issue? If so, how would you address it in your classroom?

References

Browne, N. (2004) *Gender equity in the early years*, Maidenhead: McGraw-Hill Education, Open University Press.

Browne, N. and Ross, C. (1990) 'Girls' "stuff", boys' "stuff": young children talking and playing' in Browne, N. (ed.) *Science and technology in the early years: an equal opportunities approach*, Buckingham: Open University Press.

Caleb, L. (2000) 'Design technology: learning how girls learn best', *Equity & Excellence in Education* 33(1): 22–25.

Charlton, E., Mills, M., Martino, M. and Beckett, L. (2007) 'Sacrificial girls: a case study of the impact of streaming and setting on gender reform', *British Educational Research Journal* 33(4): 459.

Cockburn, C. and Furst Dilic, R. (eds) (1994) *Bringing technology home: gender and technology in a changing Europe*, Buckingham: Open University Press.

Crawford, M. and Unger, R. (2004) *Women and gender: a feminist psychology*, 4th edn, New York: McGraw-Hill.

Curtis, P. (2009) 'Boys overtake girls in maths GCSE as coursework dropped', www.guardian.co.uk (accessed 27 August 2009).

Dakers, J.R. and Dow, W. (2009) *Understanding and providing a developmental approach to technology education: a handbook for teachers*, Glasgow: University of Glasgow.

Dakers, J., Dow, W. and McNamee, L. (2009) 'De-constructing technology's masculinity: Discovering a missing pedagogy in teaching technology', *International Journal of Design Education* 19: 381–391.

Design and Technology Association (DATA) (2011) *DATA News* 4, January 2011.

Elwood, J. (1999) 'Equity issues in performance assessment: the contribution of teacher-assessed coursework to gender-related differences in examination performance', *Educational Research and Evaluation* 5(4): 321–344.

Falsafi, L. (2010) *Learner identity. A sociocultural approach to how people recognize and construct themselves as learners.* PhD thesis, University of Barcelona; http://www.psyed.edu.es/prodGrintie/tesis/Falsafi_Thesis.pdf (accessed 25 May 2012).

Francis, B. (2011) 'Education and gender' in Lawson, N. and Spours, K. (eds) *Education for the good society*, London: Compass.

Galton, F. (1869) *Hereditary genius: an inquiry into its laws and consequences*, London: Macmillan/Fontana.

Gurian, M. and Stevens, K. (2010) *Boys and girls learn differently! A guide for teachers and parents*, rev. 10th anniversary edn, Chichester: John Wiley & Sons.

Harding, J. (1997) 'Gender and design and technology education', *Journal of Design and Technology Education* 2(1): 20–26.

Harding, J. and Grant, M. (1984) *Presenting design and technology to girls. Girls and technology education project report 84: 2*, London: Chelsea College.

Hill, C., Corbett, C. and St. Rose, A. (2010) *Why so few? Women in science, technology, engineering and mathematics*, Washington, DC: AAUW.

Hurst, G. and Sugden, J. (2010) 'Private and single sex schools still top A level and GCSE league tables', *The Times*, 13 January.

Ivinson, G. and Murphy, P. (2007) *Re-thinking single sex teaching*, Buckingham: Open University.

Jackson, C. (2002) 'Can single-sex classes in co-educational schools enhance the learning experiences of girls and/or boys? An exploration of pupils' perceptions', *British Educational Research Journal* 28(1): 37–48.

Kelly, A. Whyte, J. and Smail, B. (1984) *GIST: the final report*, Manchester: Department of Sociology, University of Manchester.

Kimbell, R., Stables, K., Wheeler, T. Wosniak, A. and Kelly, V. (1991) *The assessment of performance in design and technology*, London: Schools Examinations and Assessment Council.

Kirkup, G., Zalevski, A., Maruyama, T. and Batool, I. (2010) *Women and men in science, engineering and technology: the UK statistics guide 2010*, Bradford: UKRC.

McIntyre, R. and Woolnough, B.E. (1996) *Enriching the curriculum; evaluation report of CREST and GETSET*, London: HMSO.

Measor, L (1983) 'Gender and the sciences: pupils gender-based conceptions of school subjects' in Hammersley, M. and Hargreaves, A. (eds) *Curriculum Practice: sociological accounts*, Lewes: Falmer.

Miller, C.L. (1987) 'Qualitative differences among gender-stereotyped toys: implications for cognitive and social development in girls and boys', *Gender Roles* 16: 473–486.

Murphy, P. (2006) 'Gender and technology: gender mediation in school knowledge construction' in Dakers, J.R. (ed.) *Defining technological literacy*, New York: Palgrave Macmillan.

Newcombe, N. (1982) 'Sex-related differences in spatial ability: problems and gaps in current approaches' in Poegal, M. (ed.) *Spatial abilities: Developmental and psychological foundations*, New York: Academic Press.

OfSTED (2011) *Meeting technological challenges? Design and technology in schools 2007–10*, London: HMSO.

Parker, L.H. and Rennie, L.J. (2002) 'Teachers' implementation of gender-inclusive instructional strategies in single-sex and mixed-sex science classrooms', *International Journal of Science Education* 24(9): 881–897.

Paechter, C. (2007) *Being boys being girls: Learning masculinities and femininities*, Maidenhead: Open University Press.

Rutland, M. and Owen-Jackson, G. (2012) *Current classroom practice in the teaching of food technology: is it fit for purpose in the 21st century?* Paper presented at PATT, Stockholm, June 2012.

Sanders, J. Koch, J. and Urso, J. (1997) *Gender equity right from the start; instructional activities for teacher educators in mathematics, science and technology*, Mahwah, NJ: Lawrence Erlbaum Associates.

Shepherd, J. (2010) 'Different exams set for girls and boys', www.guardian.co.uk (accessed 30 October 2011).

Skelton, C. and Read, B. (2006) 'Male and female teachers' evaluative responses to gender and the learning environments of primary age pupils', *International Studies in Sociology of Education* 16(2): 105.

Smithers, A. (2006) 'Single-sex schools "no benefit" for girls', www.guardian.co.uk/uk/2006/jun/25/schools.gender (accessed 30 October 2011).

Smithers, A. and Robinson, P. (2006) 'The paradox of single sex schooling', Centre for Education and Employment Research, University of Buckingham Headmasters' and Headmistresses' Conference.

Stagnitti, K., Rodger, S. and Clarke, J. (1997) 'Determining gender-neutral toys for assessment of preschool children's imaginative play', *Australian Occupational Therapy Journal* 44, 119–131.

Wajcman, J. (1993) *Feminism confronts technology*, Cambridge: Polity Press.

Withey, D.R. (2000) *Opportunities for gender equality in design education*. PhD dissertation, Manchester Metropolitan University.

Withey, D.R. (2003) 'Equality issues present in teaching and workshop interaction', *Journal of Design and Technology Education* 8(1): 7–11.

Younger, M.R. and Warrington, M. (2006) 'Would Harry and Hermonie have done better in single sex classes? A review of single sex teaching in co-educational secondary schools in the United Kingdom', *American Educational Research Journal* 4: 579–620.

Chapter 13

Creativity for a new generation

David Spendlove and Alastair Wells

Introduction

We have called this chapter 'creativity for a new generation' as, instead of repeating the extensive literature on creativity, we want to offer something new. We are conscious that what we are presenting is unorthodox and challenging, looking at creativity by drawing on contemporary theories from psychology, neuroscience and philosophy. In being speculative, we want to create debate about what creativity means to teachers and the children they teach, and about the many assumptions that relate to the nurturing of creativity in design and technology (D&T).

It is often said that if you ask a class of 5-year-olds to raise their hand if they think they are creative, almost all will do so; ask the question at the end of their school life and less than half will do so. It is therefore fairly well documented that schools are not the only places in which you find out you are stupid (Holt 1995) but also where you lose your sense of optimism and perceive yourself as becoming less creative. There is an extension to this anecdote. Ask a group of student teachers at the start of their training whether they will encourage pupils to be creative and most will raise their hand. Ask the same group at the end of their training and later in their career and, again, the number of hands raised will be reduced.

These two anecdotes suggest the flame of energy, enthusiasm and optimism for creativity is somehow extinguished through the process of education, although we would hope the opposite would be true. It is not as if creativity is not valued. In 2010 IBM's Global Chief Executive Survey found that CEOs in 60 countries believe creativity is the single most important leadership quality and that creativity helps employees capitalise on complexity (IBM 2010). The report concluded that creative leaders are comfortable with ambiguity and experimentation and connect with and inspire a new generation, leading and interacting in entirely new ways. So, referring back to the two anecdotes, it would seem that students want to be creative, teachers want students to be creative and leaders of industry want students to be creative – yet the rhetoric seems to be at odds with the reality of everyday life in schools. It would seem that education itself, or to be more precise the act of 'schooling', may be at the heart of the issue and central to this chapter is the marginalising of creativity within D&T.

Creativity in design and technology

While it may be conceivable that some subjects are more creative than others it is almost inconceivable that D&T should lack creative opportunities. Yet there is consistent and extensive evidence that the subject is often bereft of creative opportunities with teachers and students often pursuing safe, formulaic approaches to their work. It is not within the remit of this chapter to explore the full extent of the reasons for this, they are already well documented (Kimbell 2000; Barlex 2003; Spendlove, 2005); however, one clear reason is fear of the unknown. Reproduction of existing privileged knowledge is much easier to conceptualise, manage, measure and be accountable for than risky creative practice. For teachers, this means that it is safer to follow well-worn paths of practice and know where you are going, even if you know that where you will end up is a pretty dull place, rather than taking students along a path that may ultimately lead nowhere but equally may lead to exciting new places.

The problem that politicians, parents and students have with creativity is that it is inherently ill-disciplined, difficult to manage, difficult to measure and difficult to conceptualise, particularly if you consider yourself as not creative. Creativity requires a leap of faith, it has to be risky and the returns from it can be low as well as high. So teachers who are highly accountable, whose reputation and performance is measured through the perceived success of their students will often, despite all their best intentions, provide their students with a benign and impoverished creative experience. Such constraining of creative opportunities in learning experiences can lead to a coercion and colluded experience (Spendlove 2010) in which students are conditioned into a response necessary for meeting a notionally 'correct' view of predetermined and reorganised knowledge consumption. This *modus operandi* has increasingly dominated the pedagogic practice of teachers, when one of our core goals as educators should be to maximise the potential for students to be creative, successful learners.

For the new teacher, such a scenario can lead to different types of behaviour. First, there are new teachers who quickly realise that providing creative opportunities may be difficult and therefore resort to mitigating against any 'creative' risks. In such cases, creative opportunities are reduced to forced choices and embellishment of ideas, such as you will produce 'X' but you may choose the colour and so on.

Second, there are new teachers who want to provide creative opportunities but are unsure how to proceed. They feel the creative risks outweigh the difficulties but are constantly frustrated because they exist in a culture that perceives performance in narrow terms.

Third, there are the teachers who are comfortable in being creative and recognising creativity in others. They can manage risk taking and recognise that the benefits far outweigh the disadvantages. They are comfortable in their teaching and place a premium on creating a creative environment.

The reality, from our experience, is that the majority of good new teachers are of the second type. They want to be encouraging and inspire through creative practice in the classroom yet recognise it challenges their pedagogic and cultural practice. This will therefore be the focus for this chapter, through three main themes:

1 unknowing to be creative
2 mirror neurons, teaching and learning
3 emotion, creativity and the teacher.

Unknowing to be creative

> We now know everything we need to know, we've invented everything that we will ever need, our job is done – grunted the caveman.

Traditional paradigms of education in D&T are predicated on a concept of robust knowledge, consistent decision making and reliable thinking, particularly in the context of designing for others and improving the quality of life for others. Our proposal offers the opposite and suggests that it may be more fruitful to develop creative approaches on the basis of recognising our cognitive limitations through a 'curriculum of unknowing'. Instead of fooling pupils into thinking how smart they, and we, are, perhaps the feature of creative education should be to show just how poor our understanding can be and that assumptions of knowing prevent the growth of creative ideas. As in our caveman example, it is tempting to think we have it all, know it all and there is nothing left to do. Yet the creative teacher and pupil both know there is much more to know, much more to find out and will continue to explore new ways of thinking.

Consider the concept of social networking. On almost every level of reliable decision making, something like Twitter should not work. The concept of 'tweeting' a message of 140 characters makes little or no sense, yet it took only 3 years, 2 months and 1 day from the first tweet ('just setting up my twttr') to the reach the billionth. Now, every week there are around a billion tweets. Setting aside the merits or otherwise of social networking it has to be acknowledged that it is a result of creative enterprise and the challenging of conventional thinking and social norms. Watts (2011) affirms this by suggesting that notional 'common sense' is a shockingly unreliable guide to being creative, yet we rely on it virtually all the time to the exclusion of other methods of thinking.

The difficulty lies in the fact that our brains are open to lots of subtle influences, which then constrain our everyday thinking and creativity. For example, if we say 'don't think about bananas' it is almost certain that you will start thinking about a banana. The example is not necessarily startling, but just planting the idea may set off a chain reaction of unconscious thinking, such as several hours later you will feel like eating a banana, you may start sketching a banana shape or you may have a short-term propensity for the colour yellow. This is what drives the creative advertising and marketing industry, so there is little new here, except that when you consider the billions of subtle factors that influence every aspect of our decision making. The difficulty with creative thinking, however, is that you are prone to start from the existing norms of everyday occurrences. To break this mould requires risky thinking, going against what you know and against instincts.

This concept of 'psychological priming' is one of many examples of our cognitive limitations, and Shubin (2008) asserts that this is a result of our convoluted evolutionary path. Far from being large-brained bipedal hominids, the end product of an evolutionary hierarchy, we are vulnerable beings susceptible to physical breakdowns and odd quirks, such as ruptured disks, hernias, hiccoughs and cancers. While the physical limitations of evolutionary development are increasingly acknowledged, the limitations of our cognitive faculties are less prone to interrogation. Yet, far from evolving as thinking machines (Stanovich 2004), we, in fact, lack autonomy of thought and are shaped by a range of cultural and biological preconditioning that shapes, hinders and constrains our everyday 'creative' thinking. Such views are now considered within the field of evolutionary psychology, which hypothesises that in the same way our bodies have evolved with built-in limitations so too has our capacity for thinking.

Sawyer (2006) identifies that not only do individuals get locked into single-minded views, but also we reinforce these views for each other until the culture itself suffers the same mindlessness. Munhall (1993) has described that moving to a state of 'unknowing' is an art and in the context of medicine proposed that the presumption of knowing leads to closure based on confidence in one's own interpretation. Therefore in medicine, adopting the state of unknowing is recognition of one's own limitations and assumptions to avoid one's own subjectivity. Such unknowing represents the intersubjective space encouraging transparency and openness of different people and different cultures to be explored. Therefore, while our brains are satisfied with knowing what they know and we often crave to be educated and learn new things, we are equally happy to operate with the often limited knowledge that we have.

Humans are also good at using the limited information they have to think creatively, but this is also a constraint as the creativity can be used as a means to avoid calibrating limited knowledge against what there might be available relative to what we know. Indeed, the model of designing adopted in many schools validates operating on such limited understanding thus promoting an illusion of knowing and reinforces the continuum of coercion. Yet, the process of designing and decision making for other people (often the focus of school D&T) would require that such thinking be optimised for best results.

The concept of unknowing is therefore offered as an antidote to the illusionary default mode of knowing. Such a position challenges the status of teachers as 'knowers' and devalues the enculturation model of knowledge accumulation and validation of knowing through public examination. Unknowing challenges the status of the school as a paradigm of certainty, knowledge assurance and reliability of decision making through acknowledgement of uncertainty. Unknowing equally represents a diversity of accomplishment and demerits collusion and coercion as the routes to progression and achievement.

We therefore argue that the first of our three themes, valuing uncertainty and recognising that all knowledge is susceptible and far from absolute, is essential in moving towards nurturing a creative environment.

Mirror neurons, teaching and learning

The second theme for developing a creative learning environment in D&T concerns the relatively recent discovery of mirror neurons as a response mechanism of human function and opens an opportunity for research in an educational setting (such as considering the nurturing of creative potential in school children). Originally discovered in the brain of macaque monkeys, mirror neurons fire both when an individual executes a particular action and when they observe another individual performing that same action (Rizzolatti et al. 1996). For example, if a monkey observes a human picking up an object, their own neural activity reacts as if they were carrying out this action themselves, even though they are not moving. Since this initial discovery, mirror neurons have been associated with many functions in humans, including empathy, action understanding and language acquisition (Blakemore and Decety 2001; Carr et al. 2003; Rizzolatti 2005). Activity in motor neurons has also been identified when people not only carry out an action but also observe an action, imagine an action (Johnson-Frey et al. 2003) or even listen to sentences describing an action (Tettamanti et al. 2005).

Design and technology aspires to engage students in creative practice, work in independent and interdependent circumstances and be innovative, but evidence from research into mirror neurons would suggest that inherent neurological structure could influence students to be more attuned to imitative learning. Adopting this theme, we offer a theoretical discussion about mirror neurons and the potential advantages of knowing what this discovery brings to reflection on pedagogical approaches to enhance creative thinking in the classroom.

There is considerable international interest in promoting *inquiry learning* as a model of education as it advocates providing students with opportunities to develop their own thinking strategies, generate ideas, think critically, identify and solve problems, and have freedom to explore and learn independently. These reforms are deemed necessary for individual welfare and economic growth and emphasise the human capacity for creativity as pivotal to the success of this initiative. Therefore exploring exciting ventures into closing the gap between neuroscience and education allows us to explore how educators (and in turn learners) could benefit from practical applications of neurological discoveries. We believe that developing an awareness of the impact of mirror neurons as a neural response mechanism can be useful for reflecting on our own practice of nurturing curiosity, creative potential, inventiveness and innovation in a D&T learning environment.

Researchers have also suggested that dysfunction of the mirror neuron system (MNS) may underlie social disorders such as autism (Dapretto et al. 2005; Iacoboni and Dapretto 2006), and data from several neurological studies support this, finding structural and functional brain differences in the MNS regions between patients with autism and those without (Hadjikhani et al. 2006; Williams et al. 2006).

There has also been a move towards further exploring the significance of mirror neurons within education, specifically in creative subjects such as art and music (Jeffers 2008; Gadberry 2010) as well as general teaching practice (Frew and Vallance 2007). In light of these new understandings, we propose to broaden the horizons further and consider the potential connections between the MNS and creativity, emotion and consciousness, exploring the underlying cognitive processes that occur as students function within an inquiry–learning design context such as technology education (Wells 2010) in an everyday classroom. In particular, we believe the connections between the MNS, empathy, emotion and creativity has significant implications for pedagogy through the benefits of encouraging empathy, exposure to creativity and modelling positive emotions alongside learning. In addition to this is the notion of learning by imitation and specifically how imitative learning fits within an inquiry–learning environment. Considering that one of the fundamental aspects of inquiry learning is the notion of students' independent exploration, there appears to be a need for balance between learning through structured modelling, versus freedom for students to generate their own ideas. This is particularly relevant in terms of creativity, a skill that would seem to require a certain degree of both.

Classroom experience shows us that as students observe and interpret the modelling or demonstration provided by their teacher, a reflection of the model appears in their own work. Some may be close approximations to the model, others may bear only a slight resemblance. It is not fully understood how mirror neurons contribute to this interpretation, but it appears they enable students to engage in thinking processes similar to those being modelled. As the students observe the teacher's action, their MNS stimulates neurological activity as though they were carrying out that action themselves. It is almost as though mirror neurons act as a sort of primer, in preparation for carrying out the action independently. Therefore, when the students come to carry out a task following a modelling session, they are able to interpret the information they have seen with added neurological connections activated by the MNS. The fact that students differ in their approximation to the model may possibly be explained by differing levels of function in the MNS, a theory that has previously been discussed in relation to autism (Dapretto et al. 2005). This has important implications for the concept of modelling in teaching, not only for modelling specific actions or skills but also for explicitly modelling the process of learning; with both teacher and learners acknowledging what particular learning is happening, how and why. Although research has shown mirror neuron activation on the observation of action, it has not previously been established whether action accompanied by language is more effective at activating the mirror neuron system in the observer. If this were the case, it would imply that teachers could potentially build on the natural neurological activity and increase MN response, by accompanying action demonstration with oral descriptions. It is possible that by explicitly referring to metacognitive processes during the learning experience, mirror neuron activation could build synaptic connections in the students' brains. Eventually, it would be easier for students to apply those thinking skills independently, as their brains would have already been 'primed' for the activity.

We therefore maintain that the second theme, recognition of the concept of mirror neurons and subconscious imitation, is essential in moving towards nurturing a creative environment. Quite simply, the learner observing and empathising with her teacher modelling the very attributes she wishes to promote generates powerful creative learning.

Emotion, creativity and the teacher

The third theme concerned with promoting and engaging in creative opportunities, based on our desire to prepare students for success in a creative learning D&T environment, is through recognition of the role of emotion in learning (Spendlove 2008a). Recent advances linking neuroscience and emotion have opened many opportunities for exploring the contribution of emotion to decision making, social functioning and creativity.

Immordino Yang and Damasio (2007) describe the profound effect of emotion on a multitude of cognitive processes within the classroom, including attention, memory and decision making. They promote the need for educators to acknowledge how emotion is fundamental for the transference of skills and knowledge learned in the classroom into a real-world environment. It is commonly appreciated that learning is most successful when there is some kind of positive emotional association with the subject or task at hand. However, as Immordino Yang and Damasio (2007) discuss, this connection goes further, in that emotion is inseparable from cognitive processes and embedded in all educational environments.

The development of empathy in the creative process is seen as an important aspect of learning, both within and beyond the classroom. It plays a fundamental role in social cohesion and interaction and arguably such skills should be nurtured within D&T education environments. The presence of empathy in the classroom also appears to be important not only for social learning but also for the development of creative thinking. Anecdotal evidence suggests that making empathetic connections with others, and being open to new ideas, provides students with an increasing accessibility to and appreciation of technological artefacts and artwork from around the globe (Jeffers 2008). Jeffers argues that teachers dedicated to preparing students for the future should give a high priority to empathy by encouraging activities that build connections between students and cultural objects, as well as between students themselves. The development of empathy allows students an insight into the minds of others, which, in turn, helps them to engage in the creative process and produce something other people will engage with. In this way, the mirror neuron system could be said to provide a foundation for the development of creative skills.

To formalise this process further, we draw on research (Spendlove 2008b, 2008c) that proposes that emotion within a creative and learning D&T-oriented environment is operationalised in three interrelated domains:

the person domain
the process domain
the product domain.

Person domain

Kress (2000) argued for a curriculum for instability, where risk and uncertainty are both welcome. Without both elements education becomes oriented towards the reproduction of existing practice and defines itself as content with existing practices. The concept of risk however 'is a largely understudied construct in the educational literature' (Reio 2005: 985) yet it is generally considered to be a prerequisite of the creative process.

There is an expectation that learners engaged in a creative D&T experience will be capable of dealing with uncertainty, risk taking, reflecting on their own performance, learning in different contexts and interrogating and creating products within a creative process. While engaging in this creative process they are developing a broad repertoire of creativity, skills, knowledge and understanding that includes an understanding of designing, making, materials, systems and although this remains a highly desirable list of elements relating to capability, the personal attributes required by a learner to achieve these outcomes are hugely ambitious and emotionally challenging.

When the context of a learner's creative endeavour is an educational one, it can be further argued that the uncertainty and risk taking are doubled, as the teacher and the learner will be equally uncertain of the outcome of any given creative challenge, therefore requiring a significant emotional investment on both parts. Indeed, it can be argued that creativity can only occur in such circumstances and that uncertainty and risk taking are essential prerequisites for creativity to take place. To exist in such an uncertain state and to be willing to take risks in pursuit of 'authenticity' requires the emotional capacity to do so. Therefore by being creative, the creator is expressing a set of values and beliefs about the world.

Ultimately to be creative is an 'expression of the self' (Morgan and Averill 1992) and such expressions and convictions require an emotional capacity, self-efficacy (Bandura 1997) or 'creative self-efficacy' (Tierney and Farmer 2002).

To take risks and deal with uncertainty in order to be innovative, however, requires the management of the emotional discomfort that comes with not always knowing how to proceed. Henderson (2004) identified that inventors expressed a profound level of emotional experience as part of their creative process. Although many emotions were mentioned, the inventors spoke repeatedly and consistently about their enjoyment of innovation work. Shaw (1994) also emphasised that negative emotions are a normal part of the creative process. One theory relating to this level of emotional discomfort is proposed by Runco (1994, 1999), who identified that creative tensions can exist when one experiences the emotional discomfort of attempting to reconcile a problem.

Therefore within the person domain, it can be argued that emotion and self-esteem are inexorably intertwined within the creative process. As such full regard has to be considered in facilitating sufficient emotional underpinning that engenders a genuine spirit of uncertainty, risk taking and creative endeavour within the learner. Without an overt recognition of the place of emotion, self-understanding and self-esteem within any activity intended to develop creative capability, that learning will ultimately be inhibited and lack true effectiveness in terms of developing capacity through the nurturing of genuine creative responses.

Process domain

It is recognised that learning is a dynamic, complex and multifaceted process in which a vast array of factors have to be in position to ensure that learning is effective. While acknowledging this within the context of the second stage of the conceptual schema, the 'process' domain of learning, attention is drawn to Vygotskian principles of meaning and sense, both being tied to emotional experience and where 'emotion-infused' mental images and 'inner speech' become the learner's focus of attention (Vygotsky 1971). Within this context, two specific areas of the emotional dimension of learning are considered: first, the emotional climate of the learner and, second, the context of emotional engagement within the learning process in D&T.

Jeffrey and Woods (1997) draw attention to the need for trust in a creative classroom. They believe that the emotional climate of the classroom needs to offer each learner personal confidence and security. Ahn (2005) suggests this is partially achieved through the exemplification of teacher modelling emotional expression, reaction and regulation, whether intentional or unintentional, which teaches the learner the nature of emotions, their expressions and how to regulate negative and positive emotion. This is consistent with our discussion of mirror neurons and imitation and would be demonstrated through the teacher's modelling of their emotional capacity to deal with both uncertainty and risk, their emotional engagement with the topic and the reinforcement and nurturing of students' emotional behaviours.

Unfortunately, within many traditional classrooms insufficient attention is given to this aspect of learning and learners are often 'cognitively, emotionally, and socially dependent on their teachers who formulate the learning goals, determine which type of interaction is allowed, and generally coerce their students to adjust to the learning environment they have created' (Boekaerts 2001: 589). Research has also shown that negative emotions, such as anxiety, fear, irritation, shame and guilt, hinder learning because they temporarily narrow the scope of attention, cognition and action (Pekrun and Perry 2002).

Product domain

Within the professional world of design, there is an increasing and growing recognition for acknowledgement and awareness of the emotional dimension (Thackara 2005). Emotional ergonomics (Seymour, cited in Bennett 2003), emotional usability (Leder et al. 2004), aesthetic emotion (Kim and Yun Moon 1989), emotional products (Demirbilek and Sener 2003) and emotional design (Norman 2004) are some examples of this phenomenon. Such acknowledgments are not generating a momentum from purely commercial expectation but from increasing demand for designers to acknowledge the full environmental, social and physical impact of their products.

The term 'product' in this domain therefore deliberately aligns the 'outputs' of a creative and learning process with such thinking and intentionally associates the outcomes with physical responses, systems, services, performances, products and artefacts that may be produced and that may be available for both the creator and others to interface and engage with. In doing this, it is recognised that the output from a creative process may not always be a 'physical' product such as those listed but may be new thinking, feelings or the development of a new skill, attitude, concept or knowledge. Within the context described, such responses would only be recognised through being externalised and by the interfacing of such outcomes with others (as in user engagement with a product, system or performance).

It must, however, be further acknowledged that tensions clearly exist when focusing purely on outcomes or products (Barlex 2003; Spendlove 2005) at the expense of true engagement with a creative process. The central argument within this chapter is that for a genuine creative and learning experience resulting in a creative outcome, there has to be a significant investment on the part of the teacher to engage in such practices. Poor practice in education is often focused, for reasons of expediency, purely on the product stages of the creative process and, in doing so, bypassing the essential creative (person) and learning (process) elements, resulting in embellished, rather than creative, novel and inspiring outcomes with limited contextualised learning, emotional engagement or opportunities to engage in risk taking and uncertainty. A key feature of our view of creativity is that the creator has a responsibility to bridge the gap between the 'receiver's' emotional needs and the emotional response they generate in others through the outcomes they have created. This is best achieved by investment and engagement in each domain as described. We therefore argue that the final of our three themes, that emotion is symbiotically intertwined with creative practice and a useful way of conceiving of this is through the person, process and product domains, is essential in moving towards nurturing a creative environment.

Conclusion

We believe that having knowledge of the emerging field of neuroscience, specifically in relation to learning, can help educators make informed decisions about the most effective pedagogical practice. Our knowledge of creativity, emotion and mirror neurons is developing all the time and, although some of it is limited to theoretical predictions, this early thinking can act as a useful metaphor in our progression towards improving creative potential in the classroom. This discussion has therefore only touched on a few issues, at times tentatively, but the intention is to open doors and make connections for the reader, providing the foundations for future explorations in this innovative and exciting field.

At the start of this chapter, we said we wanted to encourage debate by examining creativity in a different way and we believe that we have achieved this. What we have presented here is, however, only a tiny fraction of what we want to share. It is also a tiny fraction of what there is to know and it is now up to you to continue your pursuit in trying to reconcile some of the tensions we have highlighted.

Finally, earlier in this chapter we indicated that Twitter had been an unparalleled phenomenon that created a new way of communicating, so to summarise this chapter we offer these three tweets:

> Tweet one: We need to recognise our cognitive limitations and have an open mind if we are to nurture creativity in D&T.
>
> Tweet two: The concept of mirror neurons tells us that if we want our students to be creative then we have to be creative. We need to walk the talk.
>
> Tweet three: Creativity is risky and we need to consider the learners' emotions in the person, process and product stages of creative D&T.

References

Ahn, Hey Jun. (2005) 'Child care teachers' strategies in children's socialisation of emotion', *Early Child Development and Care* 175: 1–49.
Bandura, A. (1997) *Self-efficacy: The exercise of control*, New York: Freeman.
Barlex, D. (2003) *Creativity in crisis, design and technology at KS3 and KS4*: DATA Research Paper 18. Wellesbourne: DATA.
Bennett, O. (2003) 'Emotional ergonomics', *Design Week*, 11 December.

Blakemore, S. and Decety, J. (2001) 'From the perception of action to the understanding of intention', *Neuroscience* 2: 561–567.

Boekaerts, M. (2001) 'Bringing about change in the classroom: strengths and weaknesses of the self-regulated learning approach. EARLI Presidential Address', *Learning and Instruction* 12(2002): 589–604.

Carr, L., Iacoboni, M., Dubeau, M., Mazziotta, J. and Lenzi, G. (2003) 'Neural mechanisms of empathy in humans: a relay from neural systems for imitation to limbic areas', *Proceedings of the National Academy of Sciences of the United States of America* 100(9): 5497.

Dapretto, M., Davies, M., Pfeifer, J., Scott, A., Sigman, M. and Bookheimer, S. (2005) 'Understanding emotions in others: mirror neuron dysfunction in children with autism spectrum disorders', *Nature Neuroscience* 9(1): 28–30.

Demirbilek, O. and Sener, B. (2003) 'Product design, semantics and emotional response', *Ergonomics* 46(13/14): 1346–1360.

Frew, J. and Vallance, M. (2007) 'Mirror neurons: the key to learning?', *Principal Matters* 73: 2–5.

Gadberry, A. (2010) 'Modeling and the mirror neuron system', *Kodaly Envoy* 37(1): 21.

Hadjikhani, N., Joseph, R., Snyder, J. and Tager-Flusberg, H. (2006) 'Anatomical differences in the mirror neuron system and social cognition network in autism', *Cerebral Cortex* 16(9): 1276.

Henderson, J. (2004) 'Product inventors and creativity: the finer dimensions of enjoyment', *Creativity Research Journal* 16(2&3): 293–312.

Holt, J. (1995) *How children fail*, rev. edn, New York: Da Capo Press.

Iacoboni, M. and Dapretto, M. (2006) 'The mirror neuron system and the consequences of its dysfunction', *Neuroscience* 7: 942–951.

IBM (2010) *Capitalizing on complexity: insights from the Global Chief Executive Officer Study*, New York: IBM Global Business Services.

Immordino Yang, M. and Damasio, A. (2007) 'We feel, therefore we learn: the relevance of affective and social neuroscience to education', *Mind, Brain, and Education* 1(1): 3–10.

Jeffers, C. (2008) 'Empathy, cultural art, and mirror neurons: implications for the classroom and beyond', *Journal of Cultural Research in Art Education* 26: 65–71.

Jeffrey, R. and Woods, P. (1997) 'The relevance of creative teaching: pupils views' in Pollard, A., Thiessen, D., and Filer, A. (eds) *Children and their curriculum*, London: Falmer.

Johnson-Frey, S., Maloof, F., Newman-Norlund, R., Farrer, C., Inati, S. and Grafton, S. (2003) 'Actions or hand-object interactions? Human inferior frontal cortex and action observation', *Neuron* 39(6): 1053–1058.

Kim, J. and Yun Moon, J. (1989) 'Designing towards emotional usability in customer interfaces trustworthiness of cyber-banking system interfaces', *Interacting with Computers* 10: 1–29.

Kimbell, R. (2000) 'Creativity in crisis', *Journal of Design and Technology Education* 5(3): 206–211.

Kress, G. (2000) 'A curriculum for the future', *Cambridge Journal of Education* 30(1): 133–145.

Leder, H., Benno, B., Oeberst, A. and Augustin, D. (2004) 'A model of aesthetic appreciation and aesthetic judgments', *British Journal of Psychology* 95: 489–508.

Morgan, C. and Averill, J.R. (1992) 'True feelings, the self, and authenticity: a psychosocial perspective' in Franks, D.D. and Gecas, V. (eds) *Social perspectives on emotion*, Vol. 1, Greenwich, CT: JAI Press.

Munhall P. (1993) '"Unknowing": towards another pattern of knowing in nursing', *Nursing Outlook* 41(3): 125–128.

Norman, D. (2004) *Emotional design: why we love (or hate) everyday things*, New York: Basic Books.

Pekrun, R. and Perry, R.P. (2002) 'Positive emotions in education' in Frydenberg, E. (ed.) *Beyond coping: meeting goals visions and challenges*, Oxford: Oxford University Press.

Reio, T.G. (2005) 'Emotions as a lens to explore teacher identity and change: a commentary', *Teaching and Teacher Education* 21: 985–993.

Rizzolatti, G. (2005) 'The mirror neuron system and its function in humans', *Anatomy and Embryology* 210(5): 419–421.

Rizzolatti, G., Fadiga, L., Gallese, V. and Fogassi, L. (1996) 'Premotor cortex and the recognition of motor actions', *Cognitive Brain Research* 3(2): 131–141.

Runco, M.A. (1994) 'Creativity and its discontents' in Shaw, M.P. and Runco, M.A. (eds) *Creativity and affect*, Norwood, NJ: Ablex.

Runco, M.A. (1999) 'Tension, adaptability, and creativity' in Russ, S.W. (ed.) *Affect, creative experience, and psychological adjustment*, Philadelphia: Brunner/Mazel.

Sawyer, R.K. (ed.) (2006) *Cambridge handbook of the learning sciences*, New York: Cambridge University Press.

Shaw, M.P. (1994) 'Affective components of scientific creativity' in Shaw, M.P. and Runco, M.A. (eds) *Creativity and affect*, Norwood, NJ: Ablex.

Shubin, N. (2008) *Your inner fish: a journey into the 3.5-billion-year history of the human body*, New York: Pantheon Books.

Spendlove, D. (2005) 'Creativity in education: a review', *Design and Technology Education: An International Journal* 10(2): 9–18.

Spendlove, D. (2008a) 'The locating of emotion within a creative, learning and product orientated design and technology experience: person, process, product'. *International Journal of Technology and Design Education* 18: 45–57.

Spendlove, D. (2008b) *From narcissism to altruism: towards a curriculum of undoing*. Paper presented at PATT 20 International Design & Technology Education Conference Tel-Aviv, Israel, November.

Spendlove, D. (2008c) 'Still thinking and feeling: the location of emotion in the creative and learning experience (Part 2)', *Design and Technology Education: An International Journal* 13:1.

Spendlove, D. (2010) *The illusion of knowing: towards a curriculum of unknowing*. Technology Learning and Thinking Conference. Vancouver, 16–21 June 2010.

Stanovich, K.E. (2004) *The robot's rebellion: finding meaning in the age of Darwin*, Chicago: University of Chicago Press.

Tettamanti, M., Buccino, G., Saccuman, M., Gallese, V., Danna, M. and Scifo, P. (2005) 'Listening to action-related sentences activates fronto-parietal motor circuits', *Journal of Cognitive Neuroscience* 17(2), 273–281.

Thackara, J. (2005) *In the bubble: designing in a complex world*, Cambridge, MA: MIT Press.
Tierney, P. and Farmer, S.M. (2002) 'Creative self-efficacy: its potential antecedents and relationship to creative performance', *Academy of Management Journal* 45: 1137–1148.
Vygotsky, L.S. (1971) *The psychology of art*, Cambridge, MA: MIT Press.
Watts, D J. (2011) *Everything is obvious once you know the answer: how common sense fails*, London: Atlantic Books.
Wells, A.W.J. (2010) *It happens anyway: the place of creativity and design thinking in technology education*, Technology Learning and Thinking Conference, Vancouver, 16–21 June 2010.
Williams, J., Waiter, G., Gilchrist, A., Perrett, D., Murray, A. and Whiten, A. (2006) 'Neural mechanisms of imitation and mirror neuron functioning in autistic spectrum disorder', *Neuropsychologia* 44(4): 610–621.

Chapter 14

Assessment questions

David Wooff, Dawne Bell and Gwyneth Owen-Jackson

Introduction

> The most effective teachers of design and technology (D&T) combine a range of assessment strategies as they plan the work and teach their classes to ensure that assessment is integral to their teaching and the pupils' learning.
>
> (OfSTED 2003: 3)

This quote is the opening paragraph of an OfSTED report entitled 'Good assessment practice in design and technology'. But what do we mean by 'assessment' and what is considered to be good practice of this in design and technology?

The Cambridge online dictionary (http://dictionary.cambridge.org/dictionary/british/assessment?q=assessment (accessed 19 June 2012)) defines assessment as 'when you judge or decide the amount, value, quality or importance of something', a definition you can probably agree with. However, in relation to assessment in design and technology (D&T), it does raise the question of what it is that we are judging and how we make decisions about the amount, value, quality or importance. This is the essence of the debate considered in this chapter – what are we assessing in D&T and how do we make those assessments?

This debate is relevant to all countries teaching D&T, in whatever form. The literature shows that many countries, including USA, New Zealand, Sweden, Taiwan, are engaging in debates about the why, what and how of D&T assessment.

Background

Assessment is 'high stakes' in many countries across the globe. The increasing importance given by governments to comparative assessment measures such as PISA (Programme for International Student Assessment) and TIMMS (Trends in International Mathematics and Science Study) has raised the profile of national assessment outcomes.

One of the outcomes of making assessment so 'high stakes', although probably unintended, is that teachers teach to the test and pupils learn only what is required in order to pass the test. McLaren (2007: 12) cites a range of literature, which shows that 'the requirements of summative tests and external certification assessments and examinations dominate the assessment practices of many teachers'. Ecclestone (2012: 165) describes this as teachers 'coming to see their role as a translator of official criteria' and pupils as working 'compliantly, strategically and superficially'. In D&T, this is shown in research conducted by Atkinson 20 years ago, in which she found that pupils on examination courses were constrained in their design work 'in order that they could provide evidence that would meet each unit of assessment. Creative, innovative thinking played little part in the process that they used' (Atkinson 2002: 173).

The high value placed on assessment outcomes means that we need to carefully consider why we assess, what we assess and how we assess.

Why do we assess?

Assessment serves several different purposes – it can provide the pupil and teacher with information about what has been learned, it allows schools to judge one pupil or class against others, it provides data to parents, inspectors and government about school performance and, as described already, it provides data for international comparisons.

The problem, however, is that the same measure tends to be used for a variety of purposes, irrespective of its suitability for purpose. For example, test or examination results are no longer a simple measure of individual pupil performance but are also used to make judgments about the effectiveness of individual teachers, departments and whole schools. Assessment outcomes can impact on the pay and promotion of individual teachers and the reputation and status of schools. Yet, as Mansell et al. (2009: 7) point out 'results that are fit to be used for one particular (intended) purpose may not be fit to be used for another'.

One effect of the use of examination data for measuring teacher and school effectiveness is that teaching is focused on 'drilling' pupils in the methods that will gain them most marks, rather than on supporting their learning (see McLaren 2007; Ecclestone 2012). This then impacts on pupils. Harlen and Deaken Crick (2002) found that pupils placed less value on learning for its own sake and more value on learning in order to gain high marks. So, rather than raising standards as is the intention, the misuse of assessment data may have a detrimental effect, as Mansell et al. (2009: 19) suggest: '[H]olding schools to account mainly for test and examination performance, may detract from the central purpose of any education system – to improve sustained learning and the rounded education of young people.'

This debate about the purpose of assessment is a general one, it is not specific to D&T, but as teachers we should be part of a community that is asking questions about the reliability of assessment data and the purposes to which those data are put.

We now consider the central questions of this debate, in D&T what do we assess, how do we assess and what counts as assessment evidence?

What do we assess?

For assessment to be valid it should assess what it claims to assess, but what is that in D&T? In 1983 the Assessment of Performance Unit (APU) described the purpose of D&T as the development of 'technological capability', which they defined as 'the capacity to take action to master the physical world and increase the quality of life by employing the problem-solving skills, certain knowledge about energy, materials and methods of control, and the ability to make value judgments' (APU 1983: 2). This would indicate that we need to assess pupils' knowledge in given areas and their skills in solving problems and making value judgments. Is that what we do assess?

When D&T was first proposed as a National Curriculum subject in the UK, the D&T Working Group (see Chapter 1) also suggested that the aim of the subject should be to help pupils develop 'design and technological capability'. They articulated their idea of capability as meaning that pupils would be able to:

- use existing artefacts and systems effectively
- make critical appraisals of the personal, social, economic and environmental implications of artefacts and systems
- improve and extend the uses of existing artefacts and systems
- design, make and appraise new artefacts and systems
- diagnose and rectify faults in artefacts and systems. (DES/WO 1988)

A little later, the APU D&T project team decided to reject the idea of focusing assessment on the products pupils made and to concentrate 'on the thinking and decision-making processes' (APU 1991: 20) that pupils undertook in order to produce their products. They then described D&T as 'an active study, involving the purposeful pursuit of a task to some form of resolution that results in improvement (for someone) in the made world. It is a study that is essentially procedural ... and which uses knowledge and skills as a resource for action rather than regarding them as ends in themselves' (APU 1991: 17). Kimbell, Stables and Green (1996: 28) later offered the definition that technological capability is the 'combination of ability and motivation that transcends understanding and enables creative development. It provides the bridge between what is and what might be.' Is this what we assess?

These various definitions of what D&T is about, what pupils are learning, are important because we need to understand what we are teaching – and what we want pupils to learn – in order to know what we should be assessing. These various understandings of what constitutes D&T, and D&T capability have permeated the different iterations of the National Curriculum, with references to pupils using knowledge and skills to design and make products. Discussions about the nature of D&T that have taken place over many years support this view and so, we suggest, imply that what we are teaching in D&T, and therefore what we are assessing, is pupils' knowledge (of what?), their skills (in what?) and their ability to apply their knowledge and skills to design and make (what?). The questions we have inserted indicate that assessment in D&T might be problematic if we are not sufficiently clear about what it is we are assessing.

It has been suggested by Kimbell, Stables and Green (1996) that D&T helps pupils to develop their thinking and decision-making skills, and their meta-cognitive awareness of these. They posit that the D&T 'process', which involves identifying a problem or possibility, conducting research, generating and evaluating ideas and developing, making and evaluating a potential solution, makes concrete and explicit a pupil's thinking and the decisions that were made. This seems clearer, so is this what we are assessing?

Kimbell (1997) has pointed out the difficulty in trying to assess these intangible aspects – thinking, decision making, creativity. He notes that in order to do so, the intangible becomes tangible, so design thinking and creativity is assessed through 'a portfolio' and evaluation thinking and judgment becomes 'a report'. To overcome this difficulty, the APU (1991) team developed a series of tests designed to assess pupils' conceptual understanding, procedural capability and expressive or communicative capability. They further defined these as:

- conceptual qualities – those concerned with understanding materials, aesthetics and users of products
- procedural qualities – those that recognise and understand the issues contained within a task, to develop proposals, appraise work as it develops and plan work
- communication qualities – being clear and appropriate, develop in complexity, show confidence and use technical skills to communicate.

This perhaps begins to make it a little clearer as to what we should be assessing – pupils' knowledge about materials, aesthetics and products, we might want to add to this their knowledge of equipment and processes; their ability to understand a task, generate appropriate responses, plan and evaluate their own work and their ability to effectively communicate to others what they are thinking and doing. This seems much clearer to us, although we acknowledge that it requires more detail depending on the particular area of D&T in which pupils are working. The next question, then, is how do we conduct the assessment?

How do we assess?

Design and technology capability is a complex activity, which makes it difficult to assess. As the preceding section describes, much assessment work in D&T attempts to break down the wholeness of capability into its constituent parts and assess each of the parts, before putting the parts back together again. Kimbell

(1997) refers to this as 'atomistic' assessment and offers a searing critique. He argues, and we agree, that D&T capability requires pupils to have knowledge and skills that they can use appropriately to undertake real tasks to solve real problems. This is what we should be assessing, but finding ways of doing that is difficult.

Most assessments in primary and lower secondary schools in the UK are now made against National Curriculum level descriptions.

National Curriculum levels

When the National Curriculum (DES 1988) was introduced into the UK each subject had a programme of study, describing what pupils would be taught, and attainment targets, describing what they should learn. The attainment targets were graded from level 1 to level 10, but this was later reduced to eight levels. Each of these levels had a description alongside it, which stated what pupils should know, understand and be able to do at that level and it was intended that each 'level' provided a description of what pupils were expected to achieve each year. Pupils were then to be tested at the ages of 7, 11, 14 and 16 (that is, at the end of each Key Stage in school) and schools were required to report pupils' NC levels at the end of each Key Stage. It did not take long, however, before schools were using the levels to report progress at the end of each school year, then (in D&T) at the end of each unit of work and, finally, in each lesson. This is in spite of the fact that government inspectors had reported that 'the level descriptions do not discriminate sufficiently to allow year-on-year numerical reporting' (OfSTED 2003: 6).

The use of the NC level descriptions has become even more pervasive. In a food technology lesson we observed pupils had one hour in which to make a layered dessert; they had previously planned what they were going to do but no pupil had a recipe, flow chart or time sheet. The learning outcomes, presented to the pupils at the start of the lesson, were:

All will (level 5)	make a dessert with different layers
Most will (level 6)	make a dessert showing consideration for some of the specifications for a layered dessert the product will show some understanding of batch production
Some will (level 7)	make a dessert showing consideration for all of the specification for a layered dessert the product will show application of the understanding of batch production, they will be identical

Each pupil made his or her own dessert and presented two plastic pots (intended to be batch production) to the teacher who then awarded a NC level to each pupil. Although there was a brief discussion about the award of the levels, it was quite superficial. In discussion with the teacher, it seems that she was required by the school to award grades each lesson so that pupil progress could be monitored. How or why this arbitrary award of levels indicates pupil progress was not really clear.

In many schools, in order to make this lesson-by-lesson progress easier to identify, each level has been subdivided. For example, the NC description for level 5, which in relation to making products reads:

> They work from their own detailed plans, modifying them where appropriate. They work with a range of tools, materials, equipment, components and processes with some precision. They check their work as it develops and modify their approach in the light of progress.
> (Qualifications and Curriculum Authority 2007: 58)

This has been translated by one school into:

> 5a – worked independently, solving technical problems and showing evidence of creativity
>
> 5b – worked safely using a variety of tools, equipment, ingredients and techniques, and presenting a high-quality finish
>
> 5c – worked independently following a detailed plan closely, making and suggesting modifications as required.

Pupils are then expected to use these sublevels to assess their own work and to set targets for future work. It seems to us, however, that as these apply to different aspects of the work a pupil could reasonably assess that they had actually demonstrated all three levels in one lesson!

This, we argue, distorts the levels so that they become meaningless. If a level is intended to describe a pupil's D&T capability after two years of learning (each Key Stage is two to three years) how can that equate to what the pupil can do in one lesson? It seems that the government, or OfSTED (in contrast to their report in 2003), or schools, are simply looking for and implementing 'quick fixes' in the hope that they can demonstrate that pupil attainment has been raised, without considering the consequences of such implementation.

It is to be hoped that the proposed review of the National Curriculum in England also addresses the use and place of the attainment targets and level descriptions.

Another common form of assessment now found in schools in England is 'assessment for learning'.

Assessment for learning

Historically, assessment tended to be mostly summative, that is undertaken at the end of a course of study and an assessment grade or score was awarded based on pupils' performance in the end of course tests. Final tests and examinations were seen as appropriate assessment tools to measure and assess knowledge, understanding and progress in most subject areas, although some practical subjects such as craft, design and technology, often had an element of practical assessment.

A significant change to this occurred during the early 1980s. In the United Kingdom, in upper secondary schools, new vocational courses were introduced and new examinations, the General Certificate of Secondary Education (GCSE) which replaced the GCE O Level. The new forms of vocational and GCSE examinations contained elements of coursework and/or multiple assessment points, which reduced the emphasis on summative grades and made ongoing assessment more important. Then, in 1998, Black and Wiliam published their research on assessment and introduced the concept of 'assessment for learning'.

Assessment for learning (AfL) is now so embedded in schools, both primary and secondary, that no discussion of assessment is complete without mentioning it. Assessment for learning is formative assessment, it was defined by Black et al. (2002) as 'the process of seeking and interpreting evidence for use by learners and their teachers to decide where the learners are in their learning, where they need to go and how best to get there'. The principles of AfL are based on research undertaken by Black and Wiliam and the Assessment Reform Group. This showed that improving informal assessment in the classroom helped to raise pupil achievement. The link between AfL and raising pupil achievement meant that governments seized on this as a new initiative and heavily promoted the use of AfL, spending millions of pounds on promoting its use in schools.

There are four aspects to AfL: learning objectives, questioning, effective feedback and self- and peer assessment. Learning objectives, it is suggested, should be shared with pupils, lead into the development of success criteria to describe the learning and be used as a basis for questioning pupils on what they have learnt. Questions should be planned to help develop pupils' thinking and learning, wait time should be used to encourage reflection and thinking from pupils. Feedback, oral and written, should focus on the learning objectives, be constructive and positive, identify what the pupil has done well, what needs to be improved and how that improvement can be achieved. Self- and peer assessment should be used to encourage pupils to discuss their learning, what and how they have learned, what they think they need to improve and what they need to do next.

These are all sensible ideas, but the way in which AfL has been promoted and popularised in schools means that much of the AfL practice we have observed has been rote and formulaic with little understanding of what is being done and why. Black and Wiliam have also said that AfL, as they wrote about it, is not being properly implemented in schools (Stewart 2012). Black and his colleagues noted that effective use of AfL requires that 'each teacher finds his or her own ways of incorporating the lessons and ideas that are set out above into her or his own patterns of classroom work.' (Black et al. 2004: 20). Unfortunately, rather than educate teachers about AfL the government promotion focused on the use of identified strategies and ways of working that teachers have been encouraged, or required, to implement. For example, many schools now insist that teachers put the learning objectives on the board at the start of the lesson, although our observations have shown that often it is not learning objectives that appear but learning activities that pupils will be engaged in during the lesson. (This is problematic because pupil learning during and at the end of the lesson is often tested against these so-called learning objectives. If the learning objectives are activities then the teacher is simply monitoring what the pupil has done in the lesson, not what she or he has learnt!)

In some schools, not only are the learning objectives required to be on the board but also pupils are required to write them down at the start of the lesson – it is not clear to us what this is expected to contribute to their learning. A research project undertaken by James et al. found that although teachers were using AfL strategies in the classroom only 20 per cent of teachers were using them in a way that captured the 'spirit' of AfL (James et al. 2006). Similarly, an evaluation of the government's AfL strategy found that although there was evidence of good practice in some schools, in many it was ineffective due to teachers' lack of understanding and their mechanical approach to using the strategies (OfSTED 2008).

This is not to be critical of teachers. AfL is complex and needs time to be understood and conducted effectively. With all the other pressing demands on a teacher's time inside and outside the classroom there is often not the time available to read the research and literature on AfL, carefully plan lessons to include appropriate learning objectives, meaningful and relevant questions and effective self/peer assessment activities. Black et al. (2004) themselves identified that effective AfL requires teachers to have an understanding of learning theory, a deep understanding of pedagogical content knowledge and be open and willing to change their own practice. Where AfL practices become lists of activities that have to be included in the lesson, without the concomitant understanding of how or why, the effectiveness is compromised.

AfL can contribute much to assessment practice in D&T and, we argue, builds on practice that already exists. When pupils are engaged in D&T project work teachers often engage pupils in meaningful dialogue, which can reveal much about the pupil's thinking and learning; they ask questions which require the pupil to justify and explain, and they discuss with pupils the strengths of their work, what might be improved and what they need to do next. Pupils can be asked to give presentations about their work and lessons often include both self- and peer assessment. It takes only a small step from this for teachers to embrace the 'spirit' of AfL and to carefully plan the dialogue and questions used in the classroom, the tasks and activities pupils engage in and feedback given. Unfortunately, the demands now placed on teachers often limit the opportunities for authentic AfL practice.

So how do we assess in D&T? We turn again to the work of the APU D&T team (APU 1991).

Holistic or atomistic?

The discussion thus far indicates that teachers are assessing D&T by looking at the individual aspects of a pupil's work and attempting to make a judgment of it. These individual judgments are then aggregated to provide an overall D&T capability assessment. We would challenge, however, that assessment done this way is, in fact, assessing 'capability'.

We can use the analogy of learning to drive. There are different elements to driving – you have to learn how to steer the car, use the accelerator, brake and clutch, change gear, use the driving mirrors and you have to know the highway code and rules of the road. When learning to drive, you may focus on each of these elements at different times, but driving capability requires you to be able to do all these things together; it is no good being able to change gears effortlessly if you cannot steer the car! Your driving test assesses your capability to put all the elements together and you only pass if you demonstrate that you can.

The holistic assessment of D&T capability was advocated when the National Curriculum was first developed: 'Design and Technology activity is so integrative, the approach to the assessment of pupil performance ... should ideally be holistic' (DES/WO 1988). This was reinforced by the work of the APU D&T (APU 1991), which concluded that assessment of D&T capability was most effective when teachers first made a holistic judgment about a pupil's level of capability based on the evidence from the work produced. They also found that teachers showed high levels of confidence and reliability when making these holistic judgments. This holistic judgment is then considered in more detail in order to be able to describe what constitutes work that is excellent, good or poor. These descriptions then allow teachers to see where individual pupils have strengths and weaknesses in the various aspects that constitute D&T capability. The APU team argued that these individual descriptions of pupils' work were only meaningful once the holistic judgement had been made and that assessment could not be carried out in reverse, that is that the discriminators could not be added together to give a holistic judgement.

This seems to provide evidence that assessment is often best done by allowing teachers the responsibility to make judgments and then provide the evidence for this judgment from the pupil's work. There does, of course, need to be nationally agreed standards but, we argue, these are likely to arise from teachers' consensus on what constitutes good work in D&T rather than the mechanistic application of predetermined descriptions.

Having discussed methods of assessment we also need to consider what evidence from pupils' work we use to make our judgments.

What counts as assessment evidence?

Given that there is agreement that D&T assessment should be about assessing pupils' thinking as well as their making, how can we know what pupils are thinking?

As Kimbell (1991: 142) notes, pupils are usually asked to provide us with tangible evidence of their thinking and capabilities, with research and investigation capabilities evidenced through a folder, design thinking through drawings and evaluation through a report. The design 'portfolio', an example of a tangible product that tries to capture pupils' thinking, has been criticised for becoming a 'ritualisation' and, therefore, not properly understood or valued by pupils (Welch, Barlex and Taylor 2005). Mike Ive, who was a subject advisor and HMI Inspector for D&T, has described the work put into some pupil portfolios as 'neat nonsense' (cited in Barlex 2007: 53), referring to the way in which presentation is emphasised over content and pupils spend precious lesson time drawing and colouring borders on their A3 papers. While there is some validity in this criticism, we do have to consider how we can provide evidence for the judgements that we make about pupils' work.

Producing tangible evidence can be an important part of the process of designing and making for pupils. When we are supporting doctoral research students we always advise them to keep a journal with one side used to note a record of events and the other side to note a record of thinking. This is really important because as time passes and the student's thinking changes it is sometimes difficult to remember, or even know, how the thinking has changed. The same is true of pupils undertaking D&T work, if they have a record of their thinking and decision making they can see how this has developed as the work has progressed. The APU team developed the idea of building in to pupils' work various points at which they were required to 'pause for thought' in order to make a note of their thinking at a particular point in time; this provided evidence later for assessment judgments (APU 1991). Stables (2002: 142) also argues that making their thinking external and visible also allows pupils to engage in 'reflection and projection'. Reflection, she suggests, helps them to consider what they have done so far while projection helps them to think about what they might do, to 'think forward'.

Barlex (2007) suggests that, during their designing and making, pupils be encouraged to develop 'job bags', a device used by many professional designers. The job bag would contain a mix of items relevant to that particular project and that particular pupil, it might include drawings, 3D models, notes, photographs, videos, schedules. However, Barlex explains, it is not the items in the job bag themselves that would provide the assessment evidence but the pupil would be able to draw on any or all of these items in order to present, describe and justify the thinking and decision making that had taken place during the design and make process. He also suggests that the exploration of the thinking and decision-making process is done through the use of 'probes' at set points in the design and make journey, similar to the 'pause for thought' moments suggested by the APU team, so that 'evidence' is collected during the process and not just retrospectively at the end.

However, in addition to this tangible evidence there are many ways in which teachers collect assessment evidence during a lesson, some of which they may not even be conscious of. Conversations with pupils, or between pupils, can reveal their thinking, as can the questions they ask. Observations of pupils working, how they are using the tools and equipment, their body language and facial expressions, can show whether or not pupils understand or feel confident about what they are doing. Much of this evidence is ephemeral and teachers need to find a way to capture it: this could be through making notes, written or audio, or taking photographs or short videos. Pupils could be asked to make a note of a conversation or a question asked, this could be done on a Post-it note attached to the work, or by audio or video recording done by mobile phone or digital camera. Many pupils now include in project folders digital photographs of themselves working or development work they have done during the project. Pupils could be asked to make presentations of their work, which could be digitally recorded. The additional value of this is that pupils are providing evidence of their thinking while they are doing it rather than trying to recall their thinking at a later stage. Producing evidence in this way, as well as providing more authentic evidence, also reduces the assessment burden on pupils by making assessment part of the working process.

While teachers do have to be mindful of what is required for external bodies, there are opportunities in D&T to make the collection of assessment evidence more authentic, more meaningful and more representative of pupils' capability (Lawler 1999).

Conclusion

Assessment in D&T is important not only for the individual pupil but also for the teacher, the school and, it could be argued, for the D&T community. The assessments we make show what we value in the subject and what we think pupils learn by studying it. We therefore need to consider carefully what we assess in D&T.

We also need to consider how we assess pupils' D&T work. In 2007 McLaren (2007: 14) wrote that 'outdated and inappropriate assessment of learning regimes may be limiting teaching and learning' and we suggest that this is still the case today. The ubiquitous use of mobile technologies means that we can now use new assessment methods to capture relevant and meaningful evidence and make learning more interesting and challenging for pupils.

There are, of course, other debates about assessment in D&T that we have not considered – issues of differences in attainment based on gender, ethnicity and class, the reliability and validity of assessments, the standardisation of assessment across the different areas of D&T – but we believe that the issues we have considered are fundamental to the subject. First, we need to be clear what it is that we want pupils to learn then assessment needs to change from being a mechanistic tick-list to becoming an integral part of the teaching and learning process.

Questions

1 Thinking about what you teach, how would you define D&T capability?
2 Do you think D&T is best assessed holistically or in terms of its constituent elements?
3 Do you think the use of mobile technologies in D&T assessment practice really adds any value?

References

APU Assessment of Performance Unit (1983) *Report of the survey of design and technological activities in the school curriculum*, Nottingham: National Centre for School Technology/Trent Polytechnic.
APU Assessment of Performance Unit (1991) *The assessment of performance in design and technology. The final report of the APU design and technology project 1985–1991*, London: SEAC.
Atkinson, S. (2002) 'Does the need for high levels of performance curtail the development of creativity in design and technology project work? in Owen-Jackson, G. (ed.) *Teaching design and technology in secondary schools. A reader*, London: RoutledgeFalmer/Open University Press.
Barlex, D. (2007) 'Assessing capability in design and technology: the case for a minimally invasive approach', *Design and Technology Education: An International Journal* 12(2): 49–56.
Black, P., Harrison, C., Lee, C., Marshall, B. and Wiliam, D. (2002) *Working inside the black box: assessment for learning in the classroom*, London: RoutledgeFalmer.
Black, P., Harrison, C., Lee, C., Marshall, B. and Wiliam, D. (2004) 'Working inside the black box: assessment for learning in the classroom', *Phi Delta Kappa* 86(1): 9–21.
DES (Department of Education and Science) (1988) *Task Group on Assessment and Testing: a report*, London: DES.
DES/WO (Department of Education and Science/Welsh Office) (1988) *National Curriculum Design and Technology Working Group Interim Report*, London: DES.
Ecclestone, K. (2012) 'Assessment' in Arthur, J. and Peterson, A. (eds) *The Routledge companion to education*, Abingdon: Routledge.
Harlen W. and Deakin Crick R. (2002) 'A systematic review of the impact of summative assessment and tests on students' motivation for learning', *Research Evidence in Education Library*, London: EPPI-Centre, Social Science Research Unit, Institute of Education, University of London.
James, M., Black, P., Carmichael, P., Conner, C., Dudley, P. and Fox, A. (2006) *Learning how to learn: tools for schools*, Abingdon: Routledge.
Kimbell, R. (1991) 'Tackling technological tasks' in Woolnough, B. (ed.) *Practical science*, Buckingham: Open University Press.
Kimbell, R. (1997) *Assessing technology international trends in curriculum and assessment*, Buckingham: Open University Press.
Kimbell, R., Stables, K. and Green, R. (1996) *Understanding practice in design and technology*, Buckingham: Open University Press.

Lawler, T. (1999) 'Exposing the gender effects of design and technology project work by comparing strategies for presenting and managing pupils' work' in Roberts, P.H. and Norman, E.W.L. (eds) IDATER99: International Conference on Design and Technology Educational Research and Curriculum Development, Loughborough: University of Loughborough.

Mansell, W., James, M. and the Assessment Reform Group (2009) *Assessment in schools. Fit for purpose? A commentary by the Teaching and Learning Research Programme*, London: Economic and Social Research Council, Teaching and Learning Research Programme.

McLaren, S.V. (2007) 'An international overview of assessment issues in technology education: disentangling the influences, confusion and complexities', *Design and Technology Education: An International Journal* 12(2): 10–24,

OfSTED (2003) *Good assessment practice in design and technology*, London: HMSO.

OfSTED (2008) *Assessment for learning: the impact of national strategy support*, London: HMSO.

Qualifications and Curriculum Authority (2007) *Design and technology for Key Stage 3 programme of study and attainment target*, London: QCA.

Stables, K. (2002) 'Assessment in design and technology: authenticity and management issues' in Sayers, S., Morley, J. and Barnes, B. (eds) *Issues in design and technology teaching*, London: RoutledgeFalmer.

Stewart, W. (2012) 'Think you've implemented assessment for learning?', *Times Educational Supplement*, 13 July 2012; http://www.tes.co.uk/article.aspx?storycode=6261847 (accessed 13 July 2012).

Welch, M., Barlex, D. and Taylor, K. (2005) *I don't enjoy making the folder: secondary students' views of portfolios in technology education.* Paper presented at DATA International Research Conference, University of Wolverhampton.

Endpiece

The chapters in this book have debated a range of issues relevant to the teaching and learning of design and technology (D&T), from its introduction into the school curriculum and its growth around the globe to contemporary issues about its content and its contribution to pupils' education. It is hoped that some of these debates will have resonated with you and will encourage you on to further reading and further research.

As Wakefield and Owen-Jackson noted in Chapter 1, D&T as a school subject has a mildly turbulent history, which other chapters show still impact today. As changes in the economy affect societies around the world and social changes and technological developments impact on education, there is a constant need for D&T to continually monitor what is being taught, and how, to ensure that it provides pupils with a relevant and worthwhile education.

When it was first conceived, D&T was intended to be a radical innovation on the school curriculum with a purpose 'to prepare pupils to meet the needs of the twenty-first century; to stimulate originality, enterprise, practical capability in designing and making and the adaptability needed to cope with a rapidly changing society' (DfES/WO 1988). Where the subject is taught well it meets this brief. Teachers who are prepared to innovate and take risks can provide pupils with challenging and engaging activities that develop a wide range of knowledge and skills, as well as personal qualities. It is this teaching of D&T that needs to be championed and developed. Unfortunately, there are also many schools in which D&T is taught in a way that is formulaic, undemanding and repetitive. This is partly due to the increasing accountability culture, with teachers constantly having to demonstrate pupil progress and attainment. Of course, we want pupils to make progress in D&T and to demonstrate learning but this takes time and is not always demonstrable within a single lesson, yet teachers have to construct lessons in such a way that they can provide evidence of progress. Neither is learning in D&T always predictable, pupils may learn more or different things from those we intended and this should not be undervalued. And is it more important for pupils to be creative and try out things that may not work or to follow safe, routine procedures to ensure success? These are complex debates for

D&T teachers to continue to grapple with, and they take place within an ever changing educational landscape.

Education serves several different purposes, for society and for the individual, and D&T can make a valuable contribution to this. However, in order to do so it has to have a clarity of purpose and a certainty; teachers have to know and understand the subject in order to be able to explain and defend it to others. This knowledge and understanding comes partly from books such as this; whether you agree with the authors of these chapters or not it is hoped that they have expanded your understanding of some aspects of the subject.

What D&T also needs, however, is to be taken more seriously. It is not just a subject that teaches pupils practical skills, it is capable of developing their higher order thinking and a range of personal skills and qualities. In order to be able to demonstrate this, there needs to be more focused research, so that we can evidence clearly what pupils learn in the D&T classroom, and if this book has spurred you on to conduct research into aspects of the subject that is all to the good.

There is a need for research to provide evidence for the valuable learning that takes place in D&T so that the claims being made that D&T supports the development of a wide range of knowledge, skills and personal qualities can be clearly demonstrated. There is also a need for research that clearly sets out the epistemological roots of the subject. In England, there is currently an emphasis on education being concerned only with propositional knowledge, with practical and creative knowledge being squeezed out of the curriculum. The Design Commission undertook research into the role of 'design' in relation to the economy and the nature of design education in the UK. Their report recommended that design education in schools should focus on 'creative problem-solving projects' (Design Commission 2011: 12) that require interdisciplinary learning – surely, this is what D&T provides, so why is this not better known? We need not only to undertake research into the teaching and learning of D&T but also to ensure that the results of such research are widely reported.

Design and technology in England, as in other countries around the world, is currently under review. There are fears that it will no longer be part of the National Curriculum or that if it remains on the curriculum it will lose its importance and will not be provided with a programme of study. This is of concern because all pupils benefit from a good D&T education and when D&T is taught well it is one of the most motivating and rewarding subjects, for both pupils and teachers. It is also possible that not having a programme of study will liberate D&T, giving teachers the freedom to innovate, experiment and drive the subject forward. It was innovation such as this which provided the good practice on which D&T as a subject was founded, and such exciting teaching can still be found in some schools – but not all. What the D&T community needs to ensure is that in all schools D&T is understood and taught well and it is hoped that this book has made a contribution to making that happen.

References

Department for Education and Science/Welsh Office (DfES/WO) (1988) *National Curriculum Design and Technology Working Group Interim Report*, London: HMSO.

Design Commission (2011) *Restarting Britain: design, education and growth*, London: Design Commission.

Index

Note: Page numbers in *italics* are for tables, those in **bold** are for figures.

academic education 74
academic-vocational debate 2, 74–85, 147
aesthetic issues 16
African countries 2, 32; *see also* Botswana; South Africa
Ahn, Hey Jun 174
Aleman, G.R. 86, 89, 95
applied science 33, 35
apprenticeship 1, 7, 67–8
apprenticeship model of teacher education 94, 95
art and design 2, 11, 12, 17, 155; and textiles teaching 115, 116–17, 121, 122
assessment 3, 10, 180–92; evidence 188–90; formative 186; holistic versus atomistic 188; peer 186, 187; purposes of 181–2; self- 186, 187; summative 181, 185; *see also* coursework; examinations
assessment for learning (AfL) 185–7
Assessment and Performance Unit (APU) 160, 182, 183, 187, 188, 189
Assessment Reform Group 186
Atkinson, S. 181
attainment gap 55–6, 158

attainment targets 10, 11, 12, 16, 18, 34, 65, 90, 184, 185
Attar, D. 104
audio recording 190
Australia 31, 32, 34, 43–4, 46, 47
Autio, O. 81
autism 171, 172

Baker, K. 10, 31
Banks, F. 2, 31–48, 129, 143, 147
Bans, F. 87
Barlex, D. 66–7, 90, 103, 108, 119, 140, 143, 144, 145, 148, 189
behaviourism 89
Beilby, G. 108
Bell, D. 1, 3, 153–65, 180–92
Benson, C. 55
Bentley, M. 21
Bereiter, C. 80
Black, P. 186, 187
Board of Education 79
Boekaerts, M. 175
Bondi, L. 89
Botswana 47
boys: and D&T options 104, 105, 147, 153, 156, 157; and play 154–5; underachievement 55–6, 162
Bransford, J. 87, 89, 91, 92, 95
Breckon, A. 26
Brewer, D. 92
Briant, E. 92

bridging units 53, 58
British Nutrition Foundation (BNF) 103, 107
Browne, N. 154
business studies 11, 12, 17, 155

Caleb, L. 155
Callaghan, J. 10, 76, 79
Campbell, J. 21
capability 90, 182, 183–4, 188
capability tasks 25, 27–8
career choices 162, 163
carousel approach 23, 59, 103
case study approach 159, 161
catering 109, 157
Central Advisory Council for Education (CACE) 101, 102, 108
Champion, M. 139
Chapman, C. 140, 143, 144, 145
China 37–9, 47
class sizes 162; food technology 110
Claxton, G. 71
cognition: embodied 71, 90; situated 89
communicative capability 183
competencies 34
comprehensive schools 79
computer-aided design/manufacture (CAD/CAM) 18, 26–7, 106, 119, 125, 126, 127, 128, 129, 134, 135, 142, 160
computer numerical control (CNC) machines 126, 128
computer science 127
computer technology 3, 125–6
conceptual knowledge/understanding 69, 183
consciousness 171
constructivism 71, 89–90, 134, 161
cost of equipment 134–5
coursework 134, 158, 186
Cowell, T. 3, 139–50
Cox, G. 146
Coyne, R. 129

craft, design and technology (CDT) 1, 11–12, 16, 21–2, 23, 27, 102, 156
craft skills 1, 33, 65, 67–8, 79, 82, 130
Crawford, M. 162
creativity 3, 33, 105, 128, 162, 166–79, 183
CREativity in Science and Technology (CREST) 159
credibility, teachers 93
critical pedagogies approach 84, 159, 161
Crowther Report (1959) 102
culture, and food education 106
curriculum 10; content 12, *13*, 16–17, 17–18; structure 15; *see also* National Curriculum
Curriculum for Excellence (Scotland) 35–6
Czechoslovakia 32

Da Vinci Studio of Science and Engineering 75
Dakers, J.R. 55, 60, 155, 157, 159
Damasio, A. 172
Davies, P. 105
Davies, S. 3, 135–38
de Vries, M. 33, 45, 141
Deaken Crick, R. 181
Dearing Report 17
decision-making 163, 169, 183, 189
declarative knowledge 69
democratic participation 83
Denmark 33
Department for Education (DfE) 1, 17, 26, 71, 147
Department of Education and Science (DES) 10, 12, 70
Department for Education and Science (DfES) 24, 59, 193
design 3, 16, 23, 65, 90, 194; and food technology 102, 103, 105, 106; organisational focus 34; and STEM agenda 146–7; technology

as aid to 128–9; textiles 118, 119–20, 121, 122, 123
Design Commission 194
Design Council 16, 146
Design and Technology Association (DATA) 126, 135, 144, 157
Design and Technology Working Group 10–11, 14, 16, 182
Dewey, J. 83, 84
Dickens, C. 86
Digital D&T programme 26–7, 28
digital revolution 142
diplomas, vocational 77, 78
Doherty, C. 92
domestic science 65, 102, 116
Dow, W. 55, 60, 155
Dreyfus, H. and Dreyfus, S. 68
Dyson, J. 141, 146

e-portfolios 129
E-scape project 129
Eastern Europe 32, 33
Ecclestone, K. 181
economic downturn 46
economic instrumentalism 80
education: historical background 7–8; liberal 80, 84; purpose of 78–81; tripartite system of 8, 75
Education Act (1944) 8, 75, 79
Education Reform Act (1988) 8, 10, 11
'Education in Schools' Green paper (1977) 79
Eggleston, J. 8, 14–15, 54
electronics 2, 26, 127, 153, 154, 156, 157, 159, 161
Electronics in Schools Strategy 26, 27
11+ examination 75
Elshof, L. 33
embodied cognition 71, 90
emotion 68, 171, 172–6; negative 175
empathy 171, 172–3
employment 162

engineering 3, 26, 77, 145–6, 154, 162; *see also* STEM
Engineering Council 16, 103, 145
England 32–3, 34–5, 46, 47
English baccalaureate (Ebac) 115, 147, 148
environmental issues 16
equipment costs 134–5
ethical issues, and textiles teaching 120
Every Child Matters 108
evidence, assessment 188–90
evolutionary psychology 169
examinations 181; D&T 156, 157–8, 159; 11+ 75; vocational 77

Falsafi, L. 154
feedback 186
Fine, G. 103
Finland 47, 76, 81
food education 2, 17, 18, 101–14, 156; and health 106–8; historical context 101–4; as vocational education 108–9; *see also* catering; domestic science; food technology; home economics
food industry 102, 108–9
food technology 17, 19, 25, 101, 102, 110–11, 112, 153; and design 23–4; as exam choice 156; and gender debate 104–6, 153, 154, 156–7, 159, 161; and health and social care 77; Licence to Cook initiative 25, 28, 104, 107–8, 142; practical issues 109–10; and STEM agenda 141, 142–3
Foresight Project 108
formative assessment 186
France 34
Future Morph initiative 146

Galton, F. 154
Galton, M. 52, 56, 59, 60
Gatsby Charitable Foundation 26

GCSEs 186; vocational 77
gender 153–65; and attainment 157–8; attainment gap 55–6, 158; and electronics 153, 154, 156, 157, 159, 161; expectations 153–4; and food education 104–6, 153, 154, 157, 159, 161; and learner identity 153–5; and play 154–5; and resistant materials 153, 154, 156, 157, 159, 161; and science 106, 158–9, 162; stereotyping 161; teachers' 156, 157, 161; and textiles technology 153, 154, 156, 157, 159, 161
gender-neutral teaching 160–1
general national vocational qualifications (GNVQs) 77
German Democratic Republic 32
Germany 42–3
Gillard, D. 79
Ginestie, J. 91
Ginner, T. 31–2
girls: and D&T options 153, 156, 157, 158–9; and play 155; underachievement 158–9
Girls Entering Tomorrow's Science, Engineering and Technology (GETSET) 159
Girls into Science and Technology 158
Girls and Technology Education (GATE) 158
Goldhaber, D. 92
Gove, M. 147, 148
government policies 7, 8–19, 75–8
grammar schools 8, 75, 76, 101–2
Grant, C.A. 91
Grant, M. 158
Gray, J. 52, 56, 59, 60
Green, R. 52, 53, 56, 182, 183
Growney, C. 2, 51–63
The Guardian 92

Hague, C. 142
Hansen, R. 81
Harding, J. 158
Hardy, A. 3, 125–38
Harlen, W. 181
Harrison, M. 145
Hattie, J. 92, 93
health, and food education 106–8
Health Schools 108
health and social care 77
Henderson, J. 174
higher education institutions (HEI) 94, 95
Hill, A.M. 33
historical background 1, 2, 7–19, 75–8, 79
Holman, J. 140
home economics 11–12, 16, 23, 27, 102, 104, 105, 110, 111, 155, 156
Hope, G. 69
Horne, S. 107
housewifery/housecraft 8, 102
How, B. 136
Howard, T.C. 86, 89, 95
Howden Report (1926) 101, 102, 108
Hughes, C. 2, 3, 115–24, 153–65
humanistic approach 47
Hungary 32

IBM Global Chief Executive Survey 166
imitative learning 170, 171, 172, 174–5
Immordino Yang, M. 172
industrial practices 17
Industrial Revolution 7
industrialists 16
industry, links to 17
information and communications technology (ICT) 125, 127, 135–6, 143
information technology 10, 11, 12, 17, 23, 34, 37, 38, 41, 155

Ingold, T. 141
innovative trends 46–7
inquiry learning 170–1
instrumentalist approach 47
International Food and Travel Studio 75
international perspectives 31–48
International Technology Education Association (ITEA) 36
International Technology and Engineering Education Association (ITEEA) 36
Israel 34, 44–5, 47
Ive, M. 56, 60, 189
Ivinson, G. 159

James, M. 187
Jamie's School Dinners campaign 108
Japan 41, 47, 141
Jeffers, C. 173
Jeffrey, R. 174
job bags 189
Jonson, B. 129
junior technical schools 79
justifications for teaching technology 32–3

Keirl, S. 33, 91
Kerr, K. 107
Key Stage 3 strategy 24, 53
Kimbell, R. 18, 52, 53, 54, 56, 58, 59, 70, 141, 142, 148, 155, 160, 182, 183–4, 189
Kingsland, J.C. 8
knowledge 64, 65, 69–71, 90–1; conceptual 69, 183; declarative 69; procedural 69, 70, 71, 183; professional 86–90; propositional 69, 70, 71; qualitative 69; subject 90–2; tacit 69
Kolb, D.A. 68
Kress, G. 173

Lakes, R.D. 83
Lawson, S. 2, 101–14
Leach, J. 87
learner identity 153–5
learning: assessment for (AfL) 185–7; emotion and 68, 171, 172–6; imitative 170, 171, 172, 174–5; inquiry 170–1; objectives 186–7; person domain of 173–4; process domain of 174–5; product domain of 175–6; using technology to develop 130–4
learning cycle 68
learning environment 160
learning style 162
learning theory 89–90
learning through making 71
Let's Get Cooking campaign 108
Lewis, T. 140, 143, 144, 145
liberal perspective on education 80, 84
Licence to Cook initiative 25, 28, 104, 107–8, 142
literacy 53, 59–60; technological 82, 84
Loveless, A. 134
Lucas, B. 71
Luntley, M. 89

Maandag, D.W. 95
Mccormick, R. 14, 32, 34, 69, 70
McIntyre, R. 159
McLaren, S.V. 181, 190
McLellan, R. 144–5
making 1, 16, 17, 23, 27, 65, 66, 67, 71, 90, 105, 130; remote 129–30
Mansell, W. 181
manual work 8
manufacturing 77
Martin, M. 1, 2, 64–73
mathematics 10, 12, 17, 53, 143, 144–5, 146; *see also* STEM
mechatronics 41
metalwork 65

Middlesex University 26
Midland Studio School 75–6
Miller, C.L. 155
mirror neuron system (MNS) 170–2, 173, 174–5, 176
mobile technologies *126*, 129, 136, 160, 190
modelling in teaching 171–2, 174–5
Moon, B. 2
moral issues 16
Mottier, I. 141
Mozambique 32
Mulberg, C. 14
Munhall, P. 169
Murphy, P. 155, 157, 159

Nardi, E. 144
National Curriculum 1, 8–15, 22–4, 29, 52, 65, 77, 90, 127, 155; developments in 1990s 15–17, 103; developments since 2000 17–18; and food education 102, 104, 111; level descriptions 184–5
National Curriculum Council (NCC) 11–12, 14–15
National Curriculum Design and Technology Working Group 65
National Design and Technology Education Foundation (NDTEF) 23
national strategies 24, 53
National Strategies Unit 24
National Strategy Framework for D&T 27
nature versus nurture debate 154
Netherlands 34
neuroscience 170–2, 176
new vocationalism 79–80, 84
New Zealand 31, 47, 82
'Newley' Technology Initiatives 159
Newsome Report (1963) 102
Nicholl, B. 144–5
Norman, E. 15
Northern Ireland 35, 41, 75

Nuffield D&T project 25, 27–8
numeracy 53, 59–60

obesity 106–8
Obesity Unit 108
OfSTED 123, 129, 130, 135; on assessment strategies 180, 184, 187; on attainment gap 55–6; on carousel system 59; on food technology courses 103, 105–7, 111, 112
Oliver, J. 108
Olsen, J. 83
online technologies *126*, 136
O'Sullivan, G. 1, 2, 74–85
Owen-Jackson, G. 1, 2, 3, 7–20, 64–73, 86–97, 103, 129, 153–65, 180–92

Papert, S. 71
Park, H. 129
Parkes, Lady 10
partnership, primary–secondary 54–6, 60
'pause for thought' moments 189
Pavlova, M. 31
Pearson, F. 55
peer assessment 186, 187
performance targets, Sweden 34, 39–40
performativity culture 92
Perry, D. 70
person domain of learning 173–4
personal development 32
personal, social and health education (PSHE) 110
Petrina, S. 33
philosophy of education, teacher's 131, **134**
Pitt, J. 140, 142, 145
play, and gender 154–5
plenaries 27
portfolios 60, 183, 189; e- 129

postgraduate degree in education (PGCE) 94
practical work 2, 14, 65, 67, 129–30
Primary Framework Strategy 53
primary schools 10–11, 21–2, 52, 53, 91
primary–secondary partnership 54–5, 60
primary–secondary transition 2, 51–63
problem solving 33, 162–3
procedural knowledge/capability 69, 70, 71, 183
process domain of learning 174–5
product domain of learning 175–6
professional development 53, 60
Programme for International Student Assessment (PISA) 42–3, 180
programming 127
projection 189
propositional knowledge 69, 70, 71
psychological priming 169
pupil data, sharing of 55
purpose of D&T 14–15, 82

Qualifications and Curriculum Authority (QCA) 104, 127, 128
qualified teacher status (QTS) 94
qualitative knowledge 69
questioning 186
Quinn, B. 142

recording, audio/video 190
reflection 68, 189
regression, Year 7 58–60
remote making 129–30
resistant materials 2, 26, 77, 141, 153, 154, 156, 157, 159, 161
resource tasks 25, 27–8
resources: accessibility of 136; cost of 134–5; management of 134–6
risk taking 173, 174, 175
Roberts, J. 145
Robinson, P. 16, 102, 145, 146
Ross, C. 154

Royal Commission on Technical Instruction 79
Ruddock, J. 52, 56, 59, 60
Runco, M.A. 174
Rutland, M. 25, 28, 103, 105, 107–8, 110, 112, 142–3

Samuelson Commission (1882) 79
Sanders, J. 155
Sandholtz, J.H. 131
SATs (standard attainment tests) 59, 60
Saunders, M. 80
Sawyer, R.K. 169
schemes of work 15
School Board of London 79
School Direct programmes 94
School Direct salaried scheme 94
School Fruit and Vegetable Scheme 108
Schools Inquiry Commission (1864) 79
science 10, 12, 17, 33–4, 45, 140, 143–4, 146; applied 33, 35; and food technology 106; and gender 106, 158–9, 162; see also STEM
science and technology regional organisations (SATROs) 159
Science, Technology and Society project 143
Science Working Group 10
Scotland 35–6, 47, 75
secondary modern schools 8, 75, 102
Sector Skills Council for Science, Engineering and Mathematics (SEMTA) 146
self-assessment 186, 187
self-efficacy 174
self-esteem 174
Sex Discrimination Act (1975) 104, 153
Shaw, M.P. 174
Shepherd, J. 158
Shubin, N. 169

Shulman, L.S. 87, 89
Sigman, A. 142
single-gender teaching 161–2
situated cognition 89
sketching 128–9
skills 64, 66–8, 70–1; craft 1, 33, 65, 67–8, 79, 82
Smithers, A. 16, 102, 145, 146
social networking 168
socialisation 153
socioeconomic background 105, 162
South Africa 32, 34, 40–1
Soviet Union 32
Spendlove, D. 3, 166–79
Spens Report (1938) 8
spiritual issues 16
Stables, K. 52, 53, 54, 56, 58, 59, 182, 183, 189
standard attainment tests (SATs) 59, 60
starter activities 27
Steeg, T. 70, 128
STEM (science, technology, engineering and mathematics) agenda 3, 34, 37, 46, 118–19, 139–50, 159
stereotyping, gender 161
Steward, S. 144
strategic knowledge 69, 70–1
studio schools 75–6
subject knowledge, of teachers 90–2, 130–1
subject leaders 15
Sullivan, M. 21
summative assessment 181, 185
sustainability 3, 33
Sweden 2, 31, 32, 33, 34, 39–40, 47

tacit knowledge 69
Taylor, T. 141
Teach First programme 94–5
teacher education programmes 93–5
teachers: credibility 93; effectiveness 92–3; gender 156, 157, 161; performance 89; personal philosophy 131, **134**; professional development 53, 60; professional knowledge 86–90; subject knowledge 90–2, 130–1
teaching practices and approaches 21–30, 33–45
teaching to the test 181
teamwork 33, 163
technical drawing 65
technical education 8, 25–6, 37, 39, 147
technical schools 8, 75, 79
Technical and Vocational Education Initiative (TVEI) 76–7
technological literacy 82, 84
technology 10, 12, 17–19, 83, 155–7, 162; at primary level 10–11; purpose of, in D&T 127–30; use of (in D&T 135–138, 160; to develop learning 130–4)
technology education 10–11, 12, 17–18, 18–19; international perspectives on 31–48; see also STEM
Technology Enhancement Project (TEP) 25–6, 28
textiles 2, 18, 26, 77, 115–24, 141, 142, 156; and art and design 115, 116–17, 121, 122; in design and technology 115–16, 117, 119–20; and gender debate 153, 154, 156, 157, 159, 161; in the global economy 117–18; national importance of 118; and school curriculum 116–17; status as area of study 156; and STEM learning 118–19
textiles industry 116, 117–18, 119, 120, 122
theory 2, 14
thinking 188–9
three-part lesson 27
Tomlinson Report (2004) 78, 146

Training and Development Agency (TDA) 87
transition, primary–secondary 2, 51–63
transmission model 68, 161
Trends in International Mathematics and Science Study (TIMMS) 180
Twitter 168, 176

uncertainty 173, 174, 175
Understanding and Providing a Developmental Approach to Technology (UPDATE) project 155, 159, 161
Unger, R. 162
United States of America (USA) 8, 31, 33, 36–7, 46, 47
university technical colleges (UTCs) 76, 146–7
unknowing to be creative 168–70

values 16; in D&T 3
video recording 190
vocational–academic debate 2, 74–85, 147
vocational diplomas 77, 78
vocational education 32, 74, 76, 80–1; food education education as 108–9
vocational examinations 77
Vygotsky, L. 174

Wajcman, J. 155, 157
Wakefield, D. 1, 7–20, 21–48
Waldon, A. 58
Wales 36
Watts, D.J. 168
Webster, R. 71
Welch, M. 90
Wells, A. 3, 166–79
Wiliam, D. 186
Williams, J. 2, 31–48
Williams, P.J. 94
Williams, R. 1
Williamson, B. 142
Wiszniewski, D. 137
Withey, D.R. 159
Wolf Report (2011) 80–1
Women in Science and Engineering (WISE) 158
Woods, P. 174
woodwork 8, 65
Wooff, D. 1, 2, 3, 115–24, 180–92
Woolnough, B.E. 159
work experience 77

Year 6 and Year 7 differences 56, 57
Year 7 regression 58–60

Zanker, N. 2, 86–97
Zimbabwe 32